CONTEMPORARY'S
REAL NUMBERS
Developing Thinking Skills in Math
Estimation 2: Fractions and Percents

Allan D. Suter

Project Editor
Kathy Osmus

CB

CONTEMPORARY BOOKS

a division of NTC/CONTEMPORARY PUBLISHING GROUP
Lincolnwood, Illinois USA

ISBN: 0-8092-4212-5

Published by Contemporary Books,
a division of NTC/Contemporary Publishing Group, Inc.,
4255 West Touhy Avenue,
Lincolnwood (Chicago), Illinois, 60712-1975 U.S.A.

3 4 5 6 7 8 9 C(K) 16 15 14 13 12 11

Editorial Director
Caren Van Slyke

Editorial
Craig Bolt
Joe Carrig
Karin Evans
Ellen Frechette
Laura Larson
Steve Miller
Bonnie Needham
Robin O'Connor
Seija Suter

Editorial Production Manager
Norma Fioretti

Production Editor
Jean Farley Brown

Production Assistant
Marina Micari

Cover Design
Lois Koehler

Illustrator
Ophelia M. Chambliss-Jones

Typography
Impressions, Inc.
Madison, Wisconsin

Cover photo © by Michael Slaughter

CONTENTS

Learning About Estimation

Estimation helps you find an amount that is close to the exact answer. Estimation is useful when checking the accuracy of answers and especially when you don't need exact answers.

The glass is **about** $\frac{1}{3}$ full.

The gas tank is **nearly** half full.

You make reasonably accurate guesses in many different situations—traveling, cooking, buying—every day. Estimating often is more practical than finding the exact answer.

▶ Circle the letter of the most reasonable answer.

1 inch = 195 miles

1. On a map, if 1 inch = 195 miles, about how many miles does $\frac{1}{2}$ inch equal?

    a) 50   **b)** 100   **c)** 150

← part left

2. About what part of the pizza is left?

    a) $\frac{3}{4}$ **b)** $\frac{1}{2}$ **c)** $\frac{1}{4}$

3. About how much do you save?

Sale! $\frac{1}{3}$ Off

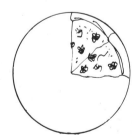

$93.99

    a) $30   **b)** $60   **c)** $90

Decide When to Estimate

In some cases, exact answers are not needed.

▶ Read each situation below and decide whether it makes more sense to estimate or to find the exact answer. Circle the letter of your answer.

1. You need to measure some boards to build a doghouse.

    a) exact   **b)** estimate

2. A group of people need to know how much pizza to order.

    a) exact   **b)** estimate

3. Marina wants just $\frac{1}{4}$ of an inch trimmed off her hair.

    a) exact   **b)** estimate

4. How much should Don leave for a tip?

    a) exact   **b)** estimate

5. Ellen needs to figure how much she earned for the year for her income taxes.

    a) exact   **b)** estimate

6. Tony has traveled half the distance of his trip.

    a) exact   **b)** estimate

Shade the Fractions

▶ Review your knowledge of fractions. Shade in the fractions below using the rectangles.

1. $\frac{3}{4}$

3 out of 4 equal parts are shaded.

2. $\frac{2}{3}$

3. $\frac{7}{8}$

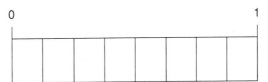

4. $\frac{1}{5}$

5. $\frac{6}{10}$

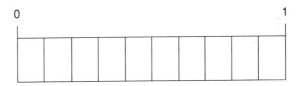

6. $\frac{3}{6}$

Estimate the Fractions

▶ Shade each rectangle to represent the fraction.

1. $\frac{4}{5}$

2. $\frac{1}{2}$

3. $\frac{2}{3}$

4. $\frac{1}{6}$

▶ Circle the fraction that shows about how much is shaded.

5. $\frac{2}{3}$ $\frac{9}{10}$ $\frac{1}{4}$

6. $\frac{1}{3}$ $\frac{3}{4}$ $\frac{5}{6}$
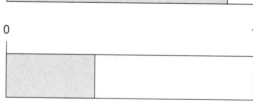

7. $\frac{1}{10}$ $\frac{1}{2}$ $\frac{3}{4}$

8. $\frac{1}{2}$ $\frac{1}{4}$ $\frac{2}{3}$
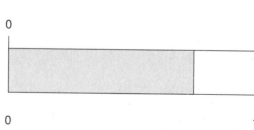

Fractions Close to 0, $\frac{1}{2}$, or 1

To estimate with fractions, it is important to know if a fraction is close to 0, $\frac{1}{2}$, or 1.

Fractions close to 0

A fraction is close to 0 when the numerator (the top number) is very small compared to the denominator (the bottom number).

Examples: $\frac{1}{16}$, $\frac{2}{15}$, $\frac{3}{20}$, $\frac{4}{100}$

Fractions close to $\frac{1}{2}$

A fraction is close to $\frac{1}{2}$ when the numerator is about half the size of the denominator.

Examples: $\frac{7}{13}$, $\frac{5}{11}$, $\frac{7}{15}$, $\frac{4}{7}$

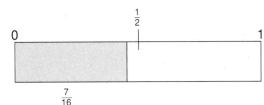

Fractions close to 1

A fraction is close to 1 when the numerator and denominator are about the same size.

Examples: $\frac{7}{8}$, $\frac{9}{10}$, $\frac{4}{5}$, $\frac{5}{6}$

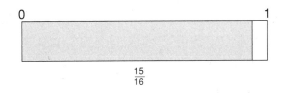

▶ Decide if each fraction is closest to 0, $\frac{1}{2}$, or 1.

1. $\frac{5}{6}$ is close to _____ .

2. $\frac{4}{9}$ is close to _____ .

3. $\frac{1}{10}$ is close to _____ .

4. $\frac{7}{8}$ is close to _____ .

5. $\frac{5}{12}$ is close to _____ .

6. $\frac{1}{20}$ is close to _____ .

Circle the Fractions

1. A fraction is close to 0 when the numerator is very small compared to the denominator. Circle the fractions that are close to 0.

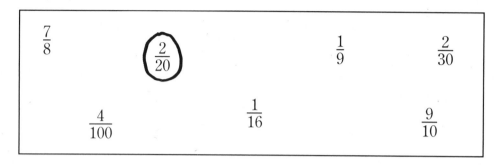

$\frac{7}{8}$ $\frac{2}{20}$ $\frac{1}{9}$ $\frac{2}{30}$

$\frac{4}{100}$ $\frac{1}{16}$ $\frac{9}{10}$

2. A fraction is close to $\frac{1}{2}$ when the numerator is about half the size of the denominator. Circle the fractions that are close to $\frac{1}{2}$.

$\frac{6}{13}$ $\frac{5}{9}$ $\frac{1}{4}$

$\frac{7}{8}$

$\frac{1}{3}$ $\frac{9}{16}$ $\frac{4}{7}$

3. A fraction is close to 1 when the numerator and denominator are about the same size. Circle the fractions that are close to 1.

$\frac{1}{4}$ $\frac{21}{23}$

$\frac{4}{5}$ $\frac{9}{10}$

$\frac{15}{16}$ $\frac{2}{5}$ $\frac{8}{9}$

Working with $\frac{1}{2}$

Sometimes when you estimate with fractions, you must decide if the fraction is greater than or less than $\frac{1}{2}$.

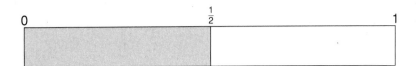

Example A
Is $\frac{5}{9}$ less than or greater than $\frac{1}{2}$?

Twice the numerator 5 is greater than the denominator 9.

$\frac{5}{9}$ is greater than $\frac{1}{2}$

Example B
Is $\frac{3}{7}$ less than or greater than $\frac{1}{2}$?

Twice the numerator 3 is less than the denominator 7.

$\frac{3}{7}$ is less than $\frac{1}{2}$

▶ Compare the fractions using the symbols < (less than) or > (greater than).

1. $\frac{7}{15}$ ⬤< $\frac{1}{2}$

Think: twice 7 is less than 15

2. $\frac{3}{8}$ ◯ $\frac{1}{2}$

3. $\frac{6}{11}$ ◯ $\frac{1}{2}$

4. $\frac{3}{5}$ ◯ $\frac{1}{2}$

5. $\frac{7}{12}$ ◯ $\frac{1}{2}$

6. $\frac{4}{9}$ ◯ $\frac{1}{2}$

7. $\frac{6}{13}$ ◯ $\frac{1}{2}$

8. $\frac{4}{6}$ ◯ $\frac{1}{2}$

Greater than or Less than $\frac{1}{2}$

 **A.** Tami made 5 out of 12 free throws. Did she make more or less than $\frac{1}{2}$ of her shots? _____
How can you tell?_____

_____

 **B.** Mario completed 9 out of 16 passes. Did he complete more or less than $\frac{1}{2}$ of his passes? _____
How can you tell?_____

_____

Sometimes when you estimate it is important to know if the fraction is greater or less than $\frac{1}{2}$.

▶ Circle the letter of the correct answer.

1. Stu answered 12 out of 25 questions correctly.

   Fraction: $\frac{12}{25}$

   a) greater than $\frac{1}{2}$
   b) less than $\frac{1}{2}$

2. Warren completed 15 out of 33 passes.

   Fraction: $\frac{15}{33}$

   a) greater than $\frac{1}{2}$
   b) less than $\frac{1}{2}$

3. Krista had 7 hits in 12 times at bat.

   Fraction: $\frac{7}{12}$

   a) greater than $\frac{1}{2}$
   b) less than $\frac{1}{2}$

4. Donna made 5 out of 9 free throws.

   Fraction: $\frac{5}{9}$

   a) greater than $\frac{1}{2}$
   b) less than $\frac{1}{2}$

5. It rained 3 out of 5 days.

   Fraction: $\frac{3}{5}$

   a) greater than $\frac{1}{2}$
   b) less than $\frac{1}{2}$

6. A rifleman hit the target 4 out of 10 times.

   Fraction: $\frac{4}{10}$

   a) greater than $\frac{1}{2}$
   b) less than $\frac{1}{2}$

Use Your Skills

1. Complete each fraction so that it equals $\frac{1}{2}$. That means the numerator will be half the size of the denominator, and the denominator will be twice the size of the numerator.

   a) $\frac{5}{10}$ b) $\frac{\square}{8}$ c) $\frac{\square}{10}$ d) $\frac{7}{\square}$ e) $\frac{3}{\square}$

2. Complete each fraction so that it is close to $\frac{1}{2}$.

   a) $\frac{\square}{9}$ b) $\frac{\square}{5}$ c) $\frac{\square}{13}$ d) $\frac{\square}{7}$ e) $\frac{\square}{11}$

3. Complete each fraction so that it is greater than $\frac{1}{2}$ and less than 1. Remember that a fraction equals 1 when the numerator and denominator are the same size.

   a) $\frac{\square}{11}$ b) $\frac{5}{\square}$ c) $\frac{4}{\square}$ d) $\frac{\square}{8}$ e) $\frac{\square}{5}$

4. Complete each fraction so that it is less than $\frac{1}{2}$.

   a) $\frac{\square}{3}$ b) $\frac{2}{\square}$ c) $\frac{5}{\square}$ d) $\frac{\square}{8}$ e) $\frac{8}{\square}$

Rounding Mixed Numbers

When estimating, sometimes it is important to round mixed numbers to whole numbers.

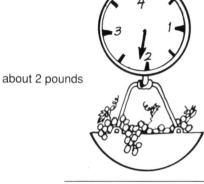

about 4 pounds

about 2 pounds

Rounding Up

Mixed numbers with fractions of $\frac{1}{2}$ or more should be rounded up to the next higher whole number.

Examples

$3\frac{3}{4}$ rounds up to 4

$7\frac{2}{3}$ rounds up to 8

Rounding Down

Mixed numbers with fractions less than $\frac{1}{2}$ should be rounded down to the next lower whole number.

Examples

$2\frac{3}{8}$ rounds down to 2

$5\frac{1}{4}$ rounds down to 5

▶ Round each mixed number to the next higher or lower whole number.

1. $1\frac{3}{8}$ rounds to _____

5. $2\frac{2}{5}$ rounds to _____

2. $9\frac{5}{6}$ rounds to _____

6. $6\frac{1}{3}$ rounds to _____

3. $7\frac{1}{4}$ rounds to _____

7. $5\frac{3}{5}$ rounds to _____

4. $4\frac{1}{2}$ rounds to _____

8. $3\frac{7}{8}$ rounds to _____

Rounding to Estimate the Sum

When adding fractions, one method to find reasonable estimates quickly is to round each mixed number to the nearest whole number and add the rounded numbers.

Example A

$3\frac{1}{8}$ rounds to   3

$+\ 4\frac{3}{4}$ rounds to $+\ 5$

Estimate: 8

Example B

$4\frac{7}{8}$ rounds to   5

$9\frac{1}{3}$ rounds to   9

$+\ 1\frac{4}{5}$ rounds to $+\ 2$

Estimate: 16

▶ Estimate the sum by rounding each mixed number.

1. $3\frac{4}{5}$ rounds to ___4___

$+\ 5\frac{1}{6}$ rounds to $+$ _____

Estimate: _____

4. $6\frac{1}{2}$ rounds to _____

$4\frac{5}{6}$ rounds to _____

$+\ 8\frac{1}{5}$ rounds to $+$ _____

Estimate: _____

2. $6\frac{1}{4}$ _____

$+\ 3\frac{3}{5}$ _____

Estimate: _____

5. $9\frac{2}{3}$ _____

$3\frac{3}{8}$ _____

$+\ 2\frac{5}{7}$ _____

Estimate: _____

3. $7\frac{3}{10}$

$+\ 2\frac{2}{3}$

Estimate: _____

6. $3\frac{5}{8}$

$8\frac{1}{2}$

$+\ 1\frac{3}{4}$

Estimate: _____

Grouping Fractions to Estimate

- Round each fraction to 0, $\frac{1}{2}$, or 1.
- Then group together the rounded fractions that equal 1.
- Then add to find the estimate.

Example A	**Example B**	**Example C**
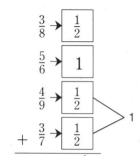	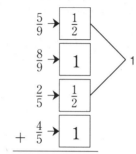	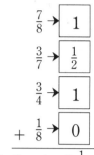
Estimate: $2\frac{1}{2}$	Estimate: 3	Estimate: $2\frac{1}{2}$

▶ Estimate the following sums by rounding each fraction and then grouping the rounded fractions to equal 1.

1.
$\frac{7}{12}$ ☐
$\frac{3}{4}$ ☐
$\frac{3}{5}$ ☐
$+ \frac{4}{9}$ ☐

Estimate: _____

3.
$\frac{8}{15}$ ☐
$\frac{7}{8}$ ☐
$\frac{11}{24}$ ☐
$+ \frac{1}{12}$ ☐

Estimate: _____

5.
$\frac{1}{3}$
$\frac{1}{4}$
$\frac{8}{9}$
$+ \frac{4}{7}$

Estimate: _____

2.
$\frac{1}{5}$ ☐
$\frac{5}{11}$ ☐
$\frac{3}{7}$ ☐
$+ \frac{11}{12}$ ☐

Estimate: _____

4.
$\frac{4}{5}$ ☐
$\frac{6}{13}$ ☐
$\frac{9}{10}$ ☐
$+ \frac{5}{6}$ ☐

Estimate: _____

6.
$\frac{3}{4}$
$\frac{2}{3}$
$\frac{3}{5}$
$+ \frac{1}{4}$

Estimate: _____

Adjusting Front-End Estimation

Another method of estimating when adding mixed numbers is to add the whole numbers and group the fractions.

Example

Step 1

Find the front-end estimate.
(Add the whole numbers.)

Step 2

Adjust the front-end estimate.

$3\frac{1}{4}$
$+ 7\frac{5}{6}$

$3\frac{1}{4}$
$+ 7\frac{5}{6}$

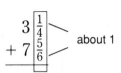

$3\,\boxed{\begin{array}{c}\frac{1}{4}\\\frac{5}{6}\end{array}}$ about 1
$+ 7$

Front-end estimate: 10

Adjusted estimate: 11

▶ First find the front-end estimate. Then group the fractions to find the adjusted estimate.

1. $\quad 4\,\boxed{\begin{array}{c}\frac{1}{8}\\\frac{3}{4}\end{array}}$ about 1
$\quad + 8$

Front-end estimate: ___12___
Adjusted estimate: _____

4. $\quad 4\frac{4}{5}$
$\quad\;\; 5\frac{3}{7}$
$\quad + 7\frac{1}{2}$

Front-end estimate: _____
Adjusted estimate: _____

2. $\quad 7\,\boxed{\begin{array}{c}\frac{2}{3}\\\frac{8}{9}\end{array}}$ about 2
$\quad + 8$

Front-end estimate: _____
Adjusted estimate: _____

5. $\quad 2\frac{2}{5}$
$\quad\;\; 8\frac{3}{5}$
$\quad + 6\frac{5}{6}$

Front-end estimate: _____
Adjusted estimate: _____

3. $\quad 9\,\boxed{\begin{array}{c}\frac{9}{10}\\\frac{1}{8}\end{array}}$ about 1
$\quad + 6$

Front-end estimate: _____
Adjusted estimate: _____

6. $\quad 4\frac{1}{3}$
$\quad\;\; 5\frac{1}{2}$
$\quad + 2\frac{7}{8}$

Front-end estimate: _____
Adjusted estimate: _____

More Adjusting Front-End Estimation

▶ Add the whole numbers and estimate the rest.

1. $3\frac{7}{8} + 8 + 2\frac{3}{8}$

    Front-end estimate: $\underline{\quad 13 \quad}$

    Adjusted estimate: $\underline{\quad 14\frac{1}{2} \quad}$

5. $7\frac{1}{3} + 2\frac{1}{6} + 1\frac{2}{5}$

    Front-end estimate: _____

    Adjusted estimate: _____

2. $4\frac{4}{5} + 7 + 3\frac{9}{10}$

    Front-end estimate: _____

    Adjusted estimate: _____

6. $8\frac{7}{8} + 9\frac{1}{3} + 1\frac{5}{6}$

    Front-end estimate: _____

    Adjusted estimate: _____

3. $9\frac{5}{8} + 7 + 2\frac{1}{3}$

    Front-end estimate: _____

    Adjusted estimate: _____

7. $2\frac{1}{4} + 6\frac{11}{12} + 3\frac{3}{8}$

    Front-end estimate: _____

    Adjusted estimate: _____

4. $1\frac{7}{16} + 5\frac{8}{9} + 5\frac{7}{8}$

    Front-end estimate: _____

    Adjusted estimate: _____

8. $1\frac{1}{2} + 5\frac{6}{7} + \frac{4}{5}$

    Front-end estimate: _____

    Adjusted estimate: _____

Estimating Larger Mixed Numbers

When adding mixed numbers greater than 10, the fractional parts of the numbers have little effect on the sum. Therefore, you can estimate by using just whole numbers.

▶ Use only the whole numbers to estimate. All reasonable answers are acceptable.

1. $21\frac{1}{5}$ → 21
    $+ 47\frac{2}{3}$ → $+ 47$

    Estimate: _____

5. $325\frac{4}{5}$ → 325
    $+ 786\frac{1}{2}$ → $+ 786$

    Estimate: _____

2. $13\frac{5}{12}$ → _____
    $+ 28\frac{7}{8}$ → $+$ _____

    Estimate: _____

6. $845\frac{2}{3}$ → _____
    $+ 315\frac{11}{12}$ → $+$ _____

    Estimate: _____

3. $16\frac{3}{4}$
    $+ 59\frac{11}{12}$

    Estimate: _____

7. $639\frac{1}{4}$
    $+ 254\frac{3}{7}$

    Estimate: _____

4. $78\frac{3}{8}$
    $+ 35\frac{1}{2}$

    Estimate: _____

8. $937\frac{1}{3}$
    $+ 85\frac{9}{16}$

    Estimate: _____

Practice Your Skills

Here are three basic methods you can use to estimate.

Method 1	Method 2	Method 3
With front-end estimation, add only the whole numbers.	Adjust the front-end estimate.	Round the mixed numbers and add.

Method 1:
$$4\tfrac{4}{5}$$
$$8\tfrac{9}{10}$$
$$+\ 1\tfrac{5}{8}$$

Estimate: 13

Method 2:
$$4\tfrac{4}{5} - \boxed{1}$$
$$8\tfrac{9}{10} - \boxed{1}$$
$$+\ 1\tfrac{5}{8} - \boxed{\tfrac{1}{2}}$$

13 and about $2\tfrac{1}{2}$ more
Adjusted estimate: $15\tfrac{1}{2}$

Method 3:
$$4\tfrac{4}{5} \text{ rounds to } 5$$
$$8\tfrac{9}{10} \text{ rounds to } 9$$
$$+\ 1\tfrac{5}{8} \text{ rounds to } +2$$

Estimate: 16

▶ Estimate the following sums using each of the three estimating methods.

1.
$$8\tfrac{7}{8}$$
$$6\tfrac{3}{5}$$
$$+\ 2\tfrac{3}{8}$$

a) Front-end: _____

b) Adjusted: _____

c) Rounded: _____

3.
$$4\tfrac{9}{10}$$
$$5\tfrac{8}{9}$$
$$+\ 3\tfrac{8}{15}$$

a) Front-end: _____

b) Adjusted: _____

c) Rounded: _____

5.
$$3\tfrac{4}{9}$$
$$9\tfrac{2}{5}$$
$$+\ 6\tfrac{5}{6}$$

a) Front-end: _____

b) Adjusted: _____

c) Rounded: _____

2.
$$5\tfrac{1}{2}$$
$$3\tfrac{1}{3}$$
$$+\ 1\tfrac{1}{8}$$

a) Front-end: _____

b) Adjusted: _____

c) Rounded: _____

4.
$$8\tfrac{12}{13}$$
$$1\tfrac{3}{4}$$
$$+\ 2\tfrac{6}{7}$$

a) Front-end: _____

b) Adjusted: _____

c) Rounded: _____

6.
$$5\tfrac{1}{6}$$
$$9\tfrac{4}{5}$$
$$7$$
$$+\ 4\tfrac{3}{4}$$

a) Front-end: _____

b) Adjusted: _____

c) Rounded: _____

Is the Answer Reasonable?

▶ Estimation is a skill we use every day. Circle the letter of the most reasonable answer.

1. Mary worked $8\frac{1}{2}$ hours yesterday and only $2\frac{1}{4}$ hours today. About how many more hours did Mary work yesterday?

    a) 10 hours   **b)** 6 hours   **c)** 1 hour

2. Rita bought $6\frac{5}{8}$ pounds of grapefruit and $10\frac{1}{3}$ pounds of apples. About how many more pounds of apples than grapefruit did she buy?

    a) 60 pounds   **b)** 17 pounds   **c)** 4 pounds

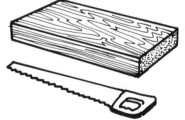

3. A board is $5\frac{1}{4}$ inches long. If the board is $2\frac{1}{2}$ inches too long, about how long must it be to fit properly?

    a) 7 inches   **b)** 3 inches   **c)** 10 inches

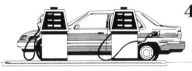

4. Mr. Niemi had $4\frac{3}{4}$ gallons of gasoline in his car. He filled up his $20\frac{1}{2}$-gallon tank. About how much gas did he buy to fill the tank?

    a) 16 gallons   **b)** 25 gallons   **c)** 80 gallons

Rounding Mixed Numbers to Subtract

One way to estimate when subtracting fractions is to round each mixed number to the nearest whole number and then subtract.

<table>
<tr><td colspan="2" align="center">Example A</td><td align="center">Example B</td></tr>
</table>

Example A

$$8\frac{1}{8} \quad \text{rounds to} \quad 8$$
$$-2\frac{3}{4} \quad \text{rounds to} \quad -3$$

Estimate: 5

Example B

$$\boxed{6} \quad \boxed{2}$$
$$5\frac{5}{8} - 2\frac{5}{12}$$
Estimate: 4

▶ Round each mixed number to the nearest whole number and subtract the rounded whole numbers for a reasonable estimate.

1. $6\frac{2}{3}$ rounds to ___7___

$-1\frac{1}{8}$ rounds to $-$ _____

Estimate: _____

2. $4\frac{3}{5}$ rounds to _____

$-2\frac{7}{8}$ rounds to $-$ _____

Estimate: _____

3. $7\frac{1}{4}$

$-4\frac{1}{6}$

Estimate: _____

4. $9\frac{2}{5}$

$-8\frac{1}{3}$

Estimate: _____

5. $\boxed{} \quad \boxed{}$

$7\frac{5}{8} - 5\frac{5}{12}$

Estimate: _____

6. $\boxed{} \quad \boxed{}$

$6\frac{5}{7} - 2\frac{3}{5}$

Estimate: _____

7. $8\frac{5}{6} - 1\frac{3}{8}$

Estimate: _____

8. $8\frac{4}{9} - 2\frac{1}{2}$

Estimate: _____

Adjust the Difference

Here is another way to estimate the difference:
- Subtract the whole numbers.
- Compare the fractional parts.

Example A

Step 1	Step 2
Subtract the whole numbers.	Compare the fractional parts.

$$5\tfrac{1}{3}$$
$$-\,2\tfrac{5}{6}$$
$$\overline{3}$$

$$5\tfrac{1}{3}$$
$$-\,2\tfrac{5}{6}$$

$\tfrac{1}{3}$ is less than $\tfrac{5}{6}$, so a closer estimate will be less than 3.

less than 3

Example B

Step 1	Step 2
Subtract the whole numbers.	Compare the fractional parts.

$$8\tfrac{7}{8}$$
$$-\,3\tfrac{1}{3}$$
$$\overline{5}$$

$$8\tfrac{7}{8}$$
$$-\,3\tfrac{1}{3}$$

$\tfrac{7}{8}$ is greater than $\tfrac{1}{3}$, so a closer estimate will be greater than 5.

greater than 5

▶ Circle the closer estimate.

1. $\quad 4\tfrac{3}{8}$
$\quad -\,1\tfrac{7}{8}$

  a) greater than 3
  b) less than 3

3. $\quad 7\tfrac{3}{7}$
$\quad -\,5\tfrac{6}{7}$

  a) greater than 2
  b) less than 2

5. $\quad 4\tfrac{3}{4}$
$\quad -\,1\tfrac{1}{2}$

  a) greater than 3
  b) less than 3

2. $\quad 7\tfrac{7}{9}$
$\quad -\,2\tfrac{1}{4}$

  a) greater than 5
  b) less than 5

4. $\quad 9\tfrac{2}{9}$
$\quad -\,2\tfrac{5}{6}$

  a) greater than 7
  b) less than 7

6. $\quad 8\tfrac{7}{12}$
$\quad -\,7\tfrac{1}{3}$

  a) greater than 1
  b) less than 1

Subtracting Larger Mixed Numbers

When subtracting mixed numbers greater than 10, the fractional parts of the numbers have little effect on the final amount. Therefore, you can estimate by using just the whole numbers.

▶ Use only the whole numbers to estimate. All reasonable answers are acceptable.

1. $47\frac{1}{6} \rightarrow \quad 47$
   $- 19\frac{3}{4} \rightarrow \quad - 19$

   Estimate: _____

5. $126\frac{1}{2} \rightarrow \quad 126$
   $- 45\frac{3}{4} \rightarrow \quad - 45$

   Estimate: _____

2. $71\frac{1}{3} \rightarrow$ _____
   $- 28\frac{7}{8} \rightarrow$ _____

   Estimate: _____

6. $304\frac{6}{7} \rightarrow$ _____
   $- 127\frac{4}{9} \rightarrow$ _____

   Estimate: _____

3. $95\frac{2}{5}$
   $- 16\frac{1}{4}$

   Estimate: _____

7. $236\frac{7}{10}$
   $- 151\frac{1}{5}$

   Estimate: _____

4. $66\frac{1}{2}$
   $- 37\frac{5}{8}$

   Estimate: _____

8. $737\frac{2}{3}$
   $- 86\frac{3}{8}$

   Estimate: _____

Estimate to Add or Subtract

▶ Take a minute to think about the numbers involved before you start.
All reasonable answers are acceptable.

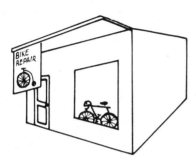

1. The bicycle repair shop opened for $4\frac{3}{4}$ hours on Monday and $8\frac{5}{6}$ hours on Tuesday. About how many more hours was the shop open on Tuesday? _____

3. Loni nailed a $3\frac{5}{8}$-foot board next to a $8\frac{1}{3}$-foot board. About what is the total length of both boards? _____

Miles Walked	
Day	**Miles**
1	$8\frac{7}{8}$
2	$5\frac{1}{4}$
3	$2\frac{2}{3}$
4	$6\frac{1}{5}$

Box	Weight (lbs.)*
1	$4\frac{1}{2}$
2	$1\frac{3}{4}$
3	$2\frac{1}{3}$
4	$8\frac{7}{8}$

* lbs. = pounds

2. a) About how many more miles were walked on Day 1 than on Day 3? _____

b) Day 3 and Day 4 totaled about how many miles? _____

c) On Day 4 about how many miles were walked? _____

4. a) Box 3 and Box 4 total about how many pounds? _____

b) About how much more does Box 4 weigh than Box 2? _____

c) The combined weight of the 4 boxes is about how much? _____

Real-World Applications

▶ Answer the following questions. Any estimate that is reasonably close to the exact answer is acceptable.

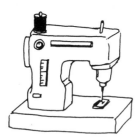

1. Lisa bought $13\frac{2}{3}$ yards of material. She used $4\frac{1}{4}$ yards to make a dress. About how much material did she have left? _____

2. Mr. Lipka drove $324\frac{1}{4}$ miles the first day and $172\frac{1}{8}$ miles the second day. About how many miles did he travel in all? _____

$\vdash\!\!\!-\!\!\!- 8\frac{1}{4}" -\!\!\!-\!\!\!\dashv\!\!\!-\, 3\frac{5}{8}" -\!\!\!\dashv$

3. A carpenter drew a line segment $8\frac{1}{4}$ inches long. He then extended the line segment $3\frac{5}{8}$ inches. About how long was the entire line segment? _____

4. Mr. Timmer had $4\frac{1}{4}$ gallons of gasoline in his car. He filled up his $19\frac{1}{2}$-gallon tank. About how much gas did he buy to fill the tank? _____

5. A baby weighed $6\frac{3}{4}$ pounds at birth. After two months she weighed $9\frac{1}{2}$ pounds. About how many pounds did she gain? _____

6. Joy bought $3\frac{5}{8}$ pounds of beef, $4\frac{1}{4}$ pounds of chicken, and a $9\frac{1}{2}$-pound ham. About how many pounds of meat did she buy? _____

Does the Answer Make Sense?

▶ Circle the letter of the most reasonable answer.

$\frac{1}{3}$ off

$196.34

1. About how much would you save?

   a) $6   **b)** $60   **c)** $90

2. If the tank holds $20\frac{1}{2}$ gallons when full, about how many gallons are left in the tank?

   a) 3   **b)** 13   **c)** 18

3. $\frac{3}{4}$ of a total of 815 students ride the bus. About how many students ride the bus?

   a) 60   **b)** 600   **c)** 800

Sale $\frac{1}{4}$ off

$13.95

4. About how much would you save?

   a) $.50   **b)** $3.50   **c)** $7.50

Rounding Mixed Numbers to Multiply

One method to find reasonable estimates when multiplying fractions is to round each mixed number to the nearest whole number and multiply the rounded whole numbers.

Example A

$$\boxed{3} \quad \boxed{5}$$

$$2\frac{3}{4} \times 5\frac{1}{3}$$

$$3 \times 5 = 15$$
estimate

Example B

$$\boxed{3} \quad \boxed{7}$$

$$3\frac{1}{8} \times 6\frac{1}{2}$$

$$3 \times 7 = 21$$
estimate

▶ Estimate by rounding each mixed number.

$$\boxed{9} \quad \boxed{5}$$

1. $8\frac{2}{3} \times 4\frac{7}{8}$

_____ × _____ = _____
estimate

5. $9\frac{1}{6} \times 4\frac{5}{8}$

_____ × _____ = _____
estimate

2. $4\frac{3}{7} \times 6\frac{3}{5}$

_____ × _____ = _____
estimate

6. $4\frac{7}{9} \times 7\frac{1}{2}$

_____ × _____ = _____
estimate

3. $1\frac{1}{8} \times 2\frac{2}{9}$

_____ × _____ = _____
estimate

7. $8\frac{3}{4} \times 2\frac{1}{6}$

_____ × _____ = _____
estimate

4. $3\frac{2}{3} \times 7\frac{1}{4}$

_____ × _____ = _____
estimate

8. $5\frac{2}{5} \times 6\frac{4}{9}$

_____ × _____ = _____
estimate

Round to Multiply

Sometimes you need to round only one of the factors.

A boat averages $6\frac{3}{4}$ miles to a gallon of gasoline. About how many miles can it travel on 9 gallons of gasoline?

$\boxed{7}$

$6\frac{3}{4} \times 9$

$7 \times 9 = 63$

estimate

The boat can travel about 63 miles on 9 gallons of gasoline.

▶ Round only the mixed number to estimate your answer.

$\boxed{3}$

1. $5 \times 3\frac{2}{5}$

_____ $\times$ _____ = $\underline{\hspace{2cm}}$

estimate

4. $8 \times 4\frac{1}{8}$

_____ $\times$ _____ = $\underline{\hspace{2cm}}$

estimate

2. $3\frac{7}{10} \times 10$

_____ $\times$ _____ = $\underline{\hspace{2cm}}$

estimate

5. $7\frac{3}{5} \times 6$

_____ $\times$ _____ = $\underline{\hspace{2cm}}$

estimate

3. $2\frac{2}{3} \times 4$

_____ $\times$ _____ = $\underline{\hspace{2cm}}$

estimate

6. $3 \times 8\frac{1}{2}$

_____ $\times$ _____ = $\underline{\hspace{2cm}}$

estimate

Rounding for Sensible Answers

Sometimes you will need to estimate mixed numbers and money amounts. Round the mixed numbers and money amounts to estimate.

About how much will you pay for $3\frac{1}{3}$ pounds of cheese?

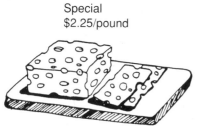

Special
$2.25/pound

$$\boxed{3} \quad \boxed{\$2}$$

$$3\frac{1}{3} \times \$2.25 = \$6.00$$

less than 5,   ↑   estimate
round down

You will pay about $6.00 for $3\frac{1}{3}$ pounds of cheese.

▶ Round the mixed numbers and money amounts to estimate.

1. About how much will you pay for $2\frac{1}{4}$ pounds of ham?

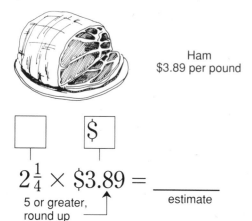

Ham
$3.89 per pound

$$\boxed{} \quad \boxed{\$}$$

$$2\frac{1}{4} \times \$3.89 = \underline{\hspace{1.5cm}}$$

5 or greater,   ↑   estimate
round up

3. About how much will you pay for $6\frac{3}{5}$ pounds of boneless chicken?

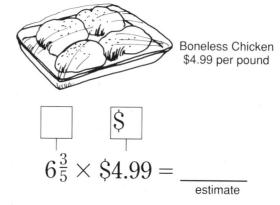

Boneless Chicken
$4.99 per pound

$$\boxed{} \quad \boxed{\$}$$

$$6\frac{3}{5} \times \$4.99 = \underline{\hspace{1.5cm}}$$

estimate

2. About how much will you pay for $5\frac{3}{4}$ pounds of peaches?

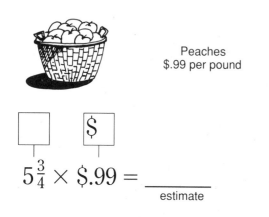

Peaches
$.99 per pound

$$\boxed{} \quad \boxed{\$}$$

$$5\frac{3}{4} \times \$.99 = \underline{\hspace{1.5cm}}$$

estimate

4. About how much will you pay for $1\frac{5}{8}$ pounds of cashews?

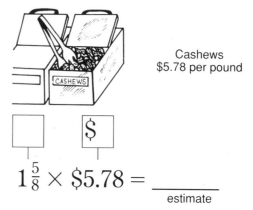

Cashews
$5.78 per pound

$$\boxed{} \quad \boxed{\$}$$

$$1\frac{5}{8} \times \$5.78 = \underline{\hspace{1.5cm}}$$

estimate

Estimating with $\frac{1}{2}$

Finding $\frac{1}{2}$ of any number is the same as dividing it by 2.

▶ Complete the following problems.

1. a) $\frac{1}{2} \times 10 =$ _____
   b) $10 \div 2 =$ _____

2. a) $\frac{1}{2} \times 6 =$ _____
   b) $6 \div 2 =$ _____

3. a) $\frac{1}{2} \times 20 =$ _____
   b) $20 \div 2 =$ _____

4. a) $\frac{1}{2} \times 50 =$ _____
   b) $50 \div 2 =$ _____

Compatible numbers are sets of numbers that are easy to figure "in your head." Finding compatible numbers is different from rounding numbers, but both methods make estimating easier.

▶ Change each dollar amount to a compatible number that makes it easy to divide by 2.

$\boxed{\$10}$

5. $\frac{1}{2} \times \$9.25 = \$$ _____
   estimate

$\boxed{\$\ }$

9. $\frac{1}{2} \times \$6.39 = \$$ _____
   estimate

Think: $18 or $20

$\boxed{\$\ }$

6. $\frac{1}{2} \times \$18.89 = \$$ _____
   estimate

$\boxed{\$\ }$

10. $\frac{1}{2} \times \$7.68 = \$$ _____
   estimate

$\boxed{\$\ }$

7. $\frac{1}{2} \times \$.94 = \$$ _____
   estimate

$\boxed{\$\ }$

11. $\frac{1}{2} \times \$8.25 = \$$ _____
   estimate

$\boxed{\$\ }$

8. $\frac{1}{2} \times \$19.98 = \$$ _____
   estimate

$\boxed{\$\ }$

12. $\frac{1}{2} \times \$13.89 = \$$ _____
   estimate

Use Compatible Numbers

Remember, when using compatible numbers to divide, think of basic division facts. Some basic division facts are:

$$5\overline{)40}^{\;8}, \quad 6\overline{)18}^{\;3}, \quad 9\overline{)36}^{\;4}, \text{ and } 4\overline{)28}^{\;7}$$

▶ Change each dollar amount to a number that is compatible with the denominator and then multiply. The dollar amount should be compatible with the denominator of the fraction given.

1. $\boxed{\$9}$ $\frac{1}{3} \times \$8.98 =$ _____
   estimate

6. $\boxed{\$20}$ $\$18.80 \times \frac{1}{5} =$ _____
   estimate

2. $\boxed{}$ $\frac{1}{4} \times \$27.45 =$ _____
   estimate

7. $\boxed{}$ $\frac{1}{10} \times \$103.55 =$ _____
   estimate

3. $\boxed{}$ $\$5.65 \times \frac{1}{2} =$ _____
   estimate

8. $\boxed{}$ $\frac{1}{3} \times \$28.19 =$ _____
   estimate

4. $\boxed{}$ $\frac{1}{4} \times \$.81 =$ _____
   estimate

9. $\boxed{}$ $\frac{1}{2} \times \$98.89 =$ _____
   estimate

5. $\boxed{}$ $\$18.76 \times \frac{1}{2} =$ _____
   estimate

10. $\boxed{}$ $\$41.80 \times \frac{1}{4} =$ _____
    estimate

Estimate What You Save

▶ Use your estimating skills to answer the following problems.

$\frac{1}{2}$ Off Sale

$19.85

1. About how much will you save by buying the sweater on sale?

_____

$\frac{1}{3}$ Off Sale

$25.10

2. About how much will you save by buying the hiking boots on sale?

_____

$\frac{1}{3}$ Off Sale

$15.49

3. About how much will you save by buying the gloves on sale?

_____

$\frac{1}{5}$ Off Sale

$33.95

4. About how much will you save by buying the calculator on sale?

_____

$\frac{1}{2}$ Off Sale

$98.89

5. About how much will you save by buying the bicycle on sale?

_____

$\frac{1}{4}$ Off Sale

$39.99

6. About how much will you save by buying the volleyball on sale?

_____

Estimate What You Would Pay

About how much would you pay if you bought the bag of cat food on sale?

$\frac{1}{4}$ Off Sale
$7.89

Step 1	Step 2	Step 3
Change each dollar amount to a compatible number.	Estimate how much you would save.	Subtract the amount you would save ($2) from the compatible number ($8).
$7.89 $\longrightarrow$ $8	$\frac{1}{4} \times \$8 = \$2$	$8 - $2 = $6
	You would save about $2.	You would pay about $6.

▶ Estimate how much you would save and pay for each item.

Item	Save	Pay
1. Box of candy $3.89 at $\frac{1}{4}$ off	_____	_____
2. Roller skates $48.95 at $\frac{1}{2}$ off	_____	_____
3. Bathing suit $26.79 at $\frac{1}{3}$ off	_____	_____
4. 10-speed bike $98.55 at $\frac{1}{5}$ off	_____	_____
5. Watch $23.98 at $\frac{1}{4}$ off	_____	_____

Figuring Costs

You can save a step in figuring costs by figuring out the fraction you will pay instead of the fraction you will save.

$\frac{1}{4}$ Off

Regular Price—$23.45

If you **save** $\frac{1}{4}$, then you must **pay** $\frac{3}{4}$.

$$\boxed{\$24}$$

Think: $\frac{3}{4} \times \$23.45 = \$18$

You would **pay** about $18 for the fishing reel.

▶ Estimate how much you would pay for each item.

$\frac{1}{4}$ Off
Regular Price
$41.68

1. If you **save** $\frac{1}{4}$, then you must **pay** $\frac{\square}{\square}$.

Think: $\frac{3}{4} \times \$40.00 =$ _____

You would pay about _____ for the clock radio.

$\frac{1}{3}$ Off
Regular Price
$88.99

2. If you **save** $\frac{\square}{\square}$, then you must **pay** $\frac{\square}{\square}$.

Think: $\frac{\square}{\square} \times$ _____ $=$ _____

You would pay about _____ for the camera.

$\frac{1}{2}$ Off
Regular Price
$17.98

3. If you **save** $\frac{\square}{\square}$, then you must **pay** $\frac{\square}{\square}$.

Think: $\frac{\square}{\square} \times$ _____ $=$ _____

You would pay about _____ for the calculator.

$\frac{1}{5}$ Off
Regular Price
$28.76

4. If you **save** $\frac{\square}{\square}$, then you must **pay** $\frac{\square}{\square}$.

Think: $\frac{\square}{\square} \times$ _____ $=$ _____

You would pay about _____ for the basketball.

Clustering

Sometimes you can cluster a group of mixed numbers around a common value. When this happens, you can arrive at a fast, reasonable estimate by using multiplication.

Monthly Rainfall	
Month	**Rainfall (in.)**
March	$2\frac{7}{8}$
April	$3\frac{1}{4}$
May	$3\frac{1}{8}$
June	$2\frac{3}{4}$

Example

Estimate the total rainfall for 4 months.

The monthly rainfall totals cluster around an average of 3 inches.

so

4×3 inches $= 12$ inches (a reasonable estimate)

▶ Use the information from the tables to answer the questions.

1. Estimate the total miles walked for 5 weeks.

   a) What common value (average) do the weekly totals cluster around? _____

   b) Estimate the answer. _____

Miles Walked	
Weeks	**Miles**
1	$9\frac{1}{2}$
2	$10\frac{3}{4}$
3	$11\frac{1}{8}$
4	$8\frac{1}{4}$
5	$10\frac{3}{8}$

2. Estimate the total weight for 6 boxes.

   a) What common value (average) do the weights cluster around? _____

   b) Estimate the answer. _____

Boxes	Weight (lbs.)
1	$4\frac{1}{2}$
2	$5\frac{3}{4}$
3	$4\frac{7}{8}$
4	5
5	$4\frac{5}{8}$
6	$5\frac{1}{4}$

Estimating to Divide

▶ Reasonable answers depend on good number sense. Circle the letter of the answer that makes the most sense.

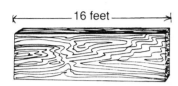

1. About how many $1\frac{3}{4}$-foot boards can be cut from a 16-foot board?

    a) 4   **b)** 8   **c)** 15

2. Helen walked $21\frac{1}{2}$ miles last week (7 days). About how many miles did she walk each day?

    a) 1   **b)** 2   **c)** 3

3. Irving rode his bicycle $18\frac{1}{2}$ miles in $2\frac{1}{2}$ hours. About how many miles did he travel in 1 hour?

    a) 9   **b)** 14   **c)** 28

4. A punch recipe calls for $\frac{2}{3}$ cup of sugar for each gallon. About how many cups of sugar are needed for $2\frac{1}{2}$ gallons?

    a) 2   **b)** 4   **c)** 8

Rounding to Estimate

When dividing fractions, a common method to find quick, reasonable estimates is to round each mixed number to the nearest whole number.

Example A

$\boxed{8}\ \ \boxed{2}$

$7\frac{5}{8} \div 1\frac{3}{4}$

$8 \div 2 = \frac{8}{2} = 4$

estimate

Example B

$\boxed{8}$

$8\frac{1}{3} \div 3$

$8 \div 3 = \frac{8}{3} = 2\frac{2}{3}$

estimate

▶ Find the estimate by rounding the mixed numbers and then dividing.

$\boxed{9}\ \ \boxed{2}$

1. $8\frac{3}{4} \div 1\frac{7}{8}$

$\square \div \square = \dfrac{\square}{\square} = \square\dfrac{\square}{\square}$

　　simplify　　estimate

5. $15\frac{3}{8} \div 4\frac{1}{9}$

$\square \div \square = \dfrac{\square}{\square} = \square\dfrac{\square}{\square}$

　　simplify　　estimate

2. $5\frac{1}{4} \div 7\frac{2}{5}$

$\square \div \square = \dfrac{\square}{\square}$

estimate

6. $9\frac{4}{5} \div 1\frac{5}{8}$

$\square \div \square = \square$

estimate

3. $8\frac{2}{3} \div 3$

$\square \div \square = \square$

7. $2\frac{2}{3} \div 5$

$\square \div \square = \dfrac{\square}{\square}$

estimate

4. $2\frac{4}{9} \div 1\frac{2}{3}$

$\square \div \square = \square$

estimate

8. $4\frac{3}{8} \div 2\frac{3}{4}$

$\square \div \square = \dfrac{\square}{\square} = \square\dfrac{\square}{\square}$

　　simplify　　estimate

Real-Life Applications

▶ Complete the following questions with reasonable estimates.

1. If a 4-gallon punch recipe calls for $6\frac{1}{8}$ cups of sugar, about how many cups of sugar are needed for 1 gallon of punch? _____

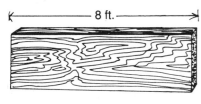

2. Mr. Ricco has a strip of wood 8 feet long. About how many $2\frac{3}{4}$-foot strips can he cut from the 8-foot strip? _____

3. Daphne rode her bicycle $15\frac{3}{4}$ miles in $2\frac{1}{4}$ hours. About how many miles did she travel in 1 hour? _____

4. Jan bought $4\frac{3}{4}$ yards of material to make 2 dresses. About how many yards of material will be used in each dress? _____

5. About how many rows $2\frac{3}{4}$ feet wide can be planted in a garden that is $11\frac{1}{2}$ feet wide? _____

Estimate the Costs

▶ Complete the following tables using information from the drawings. All reasonable estimates are acceptable.

Apples

$1.19 per pound

Shrimp

$7.84 per pound

Watermelon

$.47 per pound

Coffee

$5.39 per pound

Item	Number of Pounds Bought	Price Per Pound	Estimate the Total Cost
Watermelon	$7\frac{3}{8}$	1. a)	b)
Apples	$9\frac{1}{2}$	2. a)	b)
Coffee	$3\frac{1}{4}$	3. a)	b)
Shrimp	$2\frac{3}{4}$	4. a)	b)

Nut Mix

$6.13
per pound

Beef

$3.95
per pound

Cheese

$4.78
per pound

Broccoli

$.89
per pound

Item	Number of Pounds Bought	Price Per Pound	Estimate the Total Cost
Broccoli	$2\frac{1}{4}$	5. a)	b)
Beef	$3\frac{3}{4}$	6. a)	b)
Nut Mix	$4\frac{5}{8}$	7. a)	b)
Cheese	$1\frac{1}{8}$	8. a)	b)

Is the Answer Reasonable?

▶ Use the following drawings to estimate your answers. Decide which operation to use (addition, subtraction, multiplication, or division) before you work the problem.

$4\frac{7}{8}$ gallons

$19\frac{1}{4}$ gallons

1. If the tank was full to begin with, about how many gallons have been used? _____

2. A board is $8\frac{3}{4}$ inches long. If the board is $3\frac{1}{2}$ inches too long, about how long must it be to fit properly?

_____

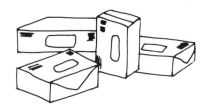

3. Ramus wants to mail 4 packages. Each package weighs $6\frac{7}{8}$ pounds. About how much do the 4 packages weigh altogether? _____

$\frac{1}{4}$ Off
Regular Price
$25.95

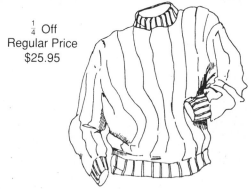

4. a) About how much will you save?

_____

b) About how much will you pay?

_____

5. Jody used $3\frac{1}{3}$ cups of sugar and $1\frac{1}{4}$ cups of milk in a recipe. About how much more sugar than milk did she use? _____

6. Juanita bought $1\frac{7}{8}$ pounds of mixed nuts and $5\frac{1}{2}$ pounds of peanuts. About how many pounds of nuts did she buy in all? _____

Review of Fraction Estimation

▶ Answer each of the following questions by estimating. Be flexible. There are many ways to estimate. Before you start a problem, take a minute to think about the numbers involved. Use a method that is quick and easy to do in your head.

1. A boat averages $9\frac{1}{4}$ miles to a gallon of gasoline. About how many miles can it travel on 21 gallons of gasoline?

_____

2. A board is $8\frac{7}{8}$ inches wide. If the board is $1\frac{3}{4}$ inches too wide, about how wide must it be to fit properly?

_____

3. Nancy walked the following number of miles each week. About how many miles did she walk after 3 weeks?

_____

Miles Walked	
Week	Miles
1	$6\frac{1}{2}$
2	$8\frac{1}{8}$
3	$19\frac{1}{4}$

4. In the problem above, about how many more miles did Nancy walk in Week 3 than in Weeks 1 and 2?

_____

5. Jesse bought $2\frac{1}{2}$ pounds of coffee at $5.99 per pound. About how much did he pay for the coffee? _____

6. Estimate about how much Mr. Waite paid for the beef and cheese altogether. _____

Item	Number of Pounds Bought	Price	Estimate
Beef	$4\frac{3}{8}$	$3.92	_____
Cheese	$1\frac{3}{4}$	$5.10	_____

7. Lynn had $6\frac{1}{4}$ gallons of gasoline in her car. She filled up her 21-gallon tank. About how much gas did she buy to fill the tank? _____

8. A 7-gallon punch recipe calls for $13\frac{1}{2}$ cups of ice cream. About how many cups of ice cream are needed for 1 gallon of punch? _____

Estimating with Percents

**Discount Air Fares
20% to 50% Off**

Census response 11.4% below goal

Response to the 1990 Census reached 62 percent, still 11.4 percent below what Census Bureau officials hope to get from the March mailing.

Pay Raise 6.5%

**SAVE
UP TO
70%
on All Merchandise!**

**20% Off
Any One Item**

**Unemployment
at 9.5%**

Percents are used in many situations every day—in business, sports, budgets, and taxes. Estimating with percents is a very important and useful life skill.

▶ Circle the letter of the most reasonable answer.

Reduced 10%

$10,495

1. About how much will you save on the car?

   a) $100   **b)** $1,000   **c)** $10,000

2. Which glass is about 50% full?

   a)   **b)** 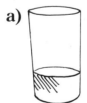   **c)**   **d)**

Was $19.98
Now $5 Off

3. About what percent will you save if you buy the basketball on sale?

   a) 25%   **b)** 50%   **c)** 75%

4. About how much would a 15% tip on this lunch bill be?

   a) $.50   **b)** $1.50   **c)** $5.50

Meaning of Percent

Percent means parts out of 100 or "per hundred." Comparing 9 to 100 can be written as $\frac{9}{100}$ or .09 (9 hundredths) or 9%.

> 50% means 50 out of every 100.

▶ Complete the following statements.

1. 30% means _____ out of every 100.

2. 10% means 10 out of every _____.

3. _____ % means 1 out of every 100.

4. 75% means _____ out of every 100.

5. $\frac{33}{100}$ = .33 = _____ %

6. 75% = ._____ = $\frac{\square}{100}$

7. $\frac{25}{\square}$ = .25 = _____ %

8. $\frac{1}{2}$ = $\frac{}{100}$ = _____ %

▶ Shade in each of the percents below.

9. 40%

0%					50%				100%

10. 91%

0%									100%

11. 35%

0%	100%

12. 59%

0%	100%

Percents Close to 100%, 10%, or 1%

To estimate with percents, it is important to know if the percent is close to 100%, 10%, or 1%.

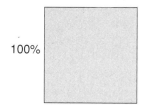

$100\% = \frac{100}{100} = 1 =$ the whole

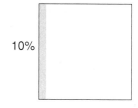

$10\% = \frac{1}{10}$ or .1 of the whole
Remember: $\frac{1}{10}$ or .1 of the whole is the same as dividing it by 10.

$1\% = \frac{1}{100}$ or .01 of the whole
Remember: $\frac{1}{100}$ or .01 of the whole is the same as dividing it by 100.

▶ Follow the pattern using 100%, 10%, and 1%.

1. 100% of 400 = <u>400</u>
    10% of 400 = <u>40</u>
    1% of 400 = <u>4</u>

3. 100% of 150 = _____
    10% of 150 = _____
    1% of 150 = _____

2. 100% of 500 = _____
    10% of 500 = _____
    1% of 500 = _____

4. 100% of 86 = _____
    10% of 86 = _____
    1% of 86 = _____

Finding 100%, 10%, and 1% of a Number

Example A (100%)	**Example B (10%)**	**Example C (1%)**
Everything	$\frac{1}{10}$	$\frac{1}{100}$
Find 100% of 275. Think: 100% is everything, so 100% of 275 = 275	Find 10% of 275. Think: 10% = $\frac{1}{10}$ or .1, so .1 × 275 = 27.5	Find 1% of 275. Think: 1% = $\frac{1}{100}$ or .01, so .01 × 275 = 2.75

▶ Circle the letter of the best answer.

1. 100% of 72 = _____   **a)** .72   **b)** 7.2   **c)** 72

2. 10% of 9 = _____   **a)** .9   **b)** 9   **c)** .09

3. 1% of 15 = _____   **a)** 15   **b)** .15   **c)** 1.5

4. 10% of 935 = _____   **a)** 9.35   **b)** 93.5   **c)** 935

5. 100% of 398 = _____   **a)** 39.8   **b)** 3.98   **c)** 398

6. 1% of 54 = _____   **a)** .54   **b)** 54   **c)** 5.4

▶ Answer with 100%, 10%, or 1%.

7. _____ % of 200 = 2

8. _____ % of 75 = 7.5

9. _____ % of 25 = 25

10. _____ % of 17 = 1.7

11. _____ % of 36 = 36

12. _____ % of 400 = 4

Using 10%

Example

Sale
20% Off
$70

How much will you save by buying the lounge chair and foot stool on sale?

Find 20% of $70.
Think: 10% of $70 = $7,
so 20% × 70 = $14

2 × 7

You will save $14.

▶ Use 10% to help you solve the following problems.

1. Find 40% of $30.
Think: $10\% \times \$30 = \$3$,
so $40\% \times \$30 =$ _____

4 × 3

5. Find 30% of $150.
Think: $10\% \times \$150 = \$15$,
so $30\% \times \$150 =$ _____

3 × 15

2. Find 60% of 90.
10% = _____
60% = _____

6. Find 20% of 300.
10% = _____
20% = _____

3. Find 90% of $20.
10% = _____
90% = _____

7. Find 70% of $80.
10% = _____
70% = _____

4. Find 80% of $50.
10% = _____
80% = _____

8. Find 20% of 350.
10% = _____
20% = _____

Using 10% to Estimate

When finding percent estimates, it is helpful to round the dollar amounts to the **lead digit** (the first digit on the left).

Sale

30% Off

$88.98

About how much will you save by buying the ring on sale?

Step 1

Round $88.98 to the lead digit.

$88.98 rounds to $90

Step 2

Find 30% of $90.
Think: $10\% \times \$90 = \$9$,
so $30\% \times \$90 = \$27$

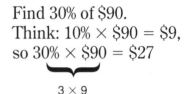

3×9

You will save about $27.

▶ Solve the following problems by rounding each dollar amount to the lead digit. Use 10% to help you estimate.

$30

1. Find 80% of $32.65.

Think: $10\% \times \$30 = $ _____ estimate

so $80\% \times \$30 = $ _____ estimate

2. Find 20% of $317.45.

$10\% = $ _____ estimate

$20\% = $ _____ estimate

3. Find 70% of $47.89.

$10\% = $ _____ estimate

$70\% = $ _____ estimate

$600

4. Find 40% of $583.00.

Think: $10\% \times \$600 = $ _____ estimate

so $40\% \times \$600 = $ _____ estimate

5. Find 90% of $74.10.

$10\% = $ _____ estimate

$90\% = $ _____ estimate

6. Find 60% of $179.45.

$10\% = $ _____ estimate

$60\% = $ _____ estimate

Using 1%

Using 1% can help you to solve problems with 2% through 9%.

Example A

Find 2% of $400.
Think: 1% × $400 = $4,
so 2% × $400 = $8

2 × $4

Example B

Find 5% of $30.
Think: 1% × $30 = $.30,
so 5% × $30 = $1.50

5 × $.30

▶ Use 1% to help you solve the following problems.

1. Find 2% of $600.

Think: 1% × $600 = $6,

so 2% × $600 = _____

2. Find 4% of $200.

1% = _____

4% = _____

3. Find 7% of $500.

1% = _____

7% = _____

4. Find 8% of $7,000.

1% = _____

8% = _____

5. Find 5% of $25.

Think: 1% × $25 = $.25,

so 5% × $25 = _____

6. Find 2% of $12.

1% = _____

2% = _____

7. Find 6% of $20.

1% = _____

6% = _____

8. Find 3% of $1,500.

1% = _____

3% = _____

Using 1% to Estimate

About how much will you save if you pay $18.98 in cash for gasoline?

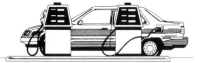

Gas for Less
Pay Cash
and
Save 3%

<u>Step 1</u>

Round $18.98 to the lead digit.

$18.98 rounds to $20

<u>Step 2</u>

$20

Find 3% of $18.98.
Think: 1% × $20 = $.20,
so 3% × $20 = $.60

You will save about $.60 by paying cash for the gasoline.

▶ Solve the following problems by rounding each dollar amount to the lead digit. Use 1% to help you estimate.

$10

1. Find 5% of $9.89.

Think: 1% × $10 = _____
       estimate

so 5% × $10 = _____
         estimate

2. Find 2% of $72.99.

1% = _____
   estimate

2% = _____
   estimate

3. Find 8% of $4,750.

1% = _____
   estimate

8% = _____
   estimate

4. Find 3% of $7.25.

1% = _____
   estimate

3% = _____
   estimate

5. Find 7% of $8,275.

1% = _____
   estimate

7% = _____
   estimate

6. Find 9% of $1,895.

1% = _____
   estimate

9% = _____
   estimate

Round to Lead Digits

Sometimes it is helpful to round both the percent and the number amount to the lead digit to estimate.

- If both numbers are rounded *up,* the estimate will be an **overestimate** (more than the actual answer).

- If both numbers are rounded *down,* the estimate will be an **underestimate** (less than the actual answer).

▶ Is your rounded estimate an overestimate or underestimate? Circle the letter of the correct choice.

1. Find 2.3% of $2,275.

1% = _____ so 2% = _____
      estimate   estimate

a) overestimate   **b)** underestimate

4. Find 17% of $495.

10% = _____ so 20% = _____
       estimate   estimate

a) overestimate   **b)** underestimate

2. Find 87% of $29.15.

10% = _____ so 90% = _____
       estimate   estimate

a) overestimate   **b)** underestimate

5. Find 1.75% of $9,750.

1% = _____ so 2% = _____
      estimate   estimate

a) overestimate   **b)** underestimate

3. Find 11.75% of $7,299.

10% = _____
       estimate

a) overestimate   **b)** underestimate

6. Find 5.15% of $54.

1% = _____ so 5% = _____
      estimate   estimate

a) overestimate   **b)** underestimate

Estimate a 15% Tip

About how much is a 15% tip on a dinner bill of $23.73?

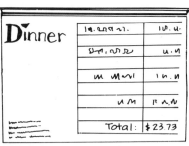

Step 1

Round $23.73 to the lead digit.

$23.73 rounds to $20

The tip will be about $3.

Step 2

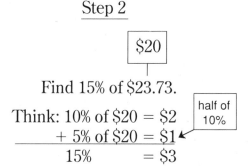

Find 15% of $23.73.

Think: 10% of $20 = $2
+ 5% of $20 = $1
15% = $3

half of 10%

▶ Round each dollar amount to the lead digit. Estimate a 15% tip.

1. Dinner bill—$38.95

$$10\% = \underline{\hspace{2cm}}$$
$$+\ 5\% = +\ \underline{\hspace{2cm}}$$
$$15\% = \underline{\hspace{2cm}}$$
estimate

half of 10%

4. Dinner bill—$76.12

$$10\% = \underline{\hspace{2cm}}$$
$$+\ 5\% = +\ \underline{\hspace{2cm}}$$
$$15\% = \underline{\hspace{2cm}}$$
estimate

2. Dinner bill—$62.15

$$10\% = \underline{\hspace{2cm}}$$
$$+\ 5\% = +\ \underline{\hspace{2cm}}$$
$$15\% = \underline{\hspace{2cm}}$$
estimate

5. Dinner bill—$4.73

$$10\% = \underline{\hspace{2cm}}$$
$$+\ 5\% = +\ \underline{\hspace{2cm}}$$
$$15\% = \underline{\hspace{2cm}}$$
estimate

3. Dinner bill—$9.53

$$10\% = \underline{\hspace{2cm}}$$
$$+\ 5\% = +\ \underline{\hspace{2cm}}$$
$$15\% = \underline{\hspace{2cm}}$$
estimate

6. Dinner bill—$97.25

$$10\% = \underline{\hspace{2cm}}$$
$$+\ 5\% = +\ \underline{\hspace{2cm}}$$
$$15\% = \underline{\hspace{2cm}}$$
estimate

Estimate the Sales Tax

The sales tax is 6% of your purchase of $432. About how much is the sales tax?

2 for
$432

$$\boxed{400}$$

Find 6% of $432.

Think: 1% × 400 = $4,
so 6% × 400 = $24
estimate

The sales tax is about $24.

▶ Estimate the sales tax for each of the following purchases and fill in the estimates in the chart below. Then show whether your answer is an overestimate or an underestimate.

	Amount of Purchase	Sales Tax Rate	Sales Tax Estimate	Circle the Answer
1.	$50 $51.50	6%	_____	a) overestimate b) underestimate
2.	[ ] $893	4%	_____	a) overestimate b) underestimate
3.	[ ] $18.32	[ ] 5.8%	_____	a) overestimate b) underestimate
4.	[ ] $193	[ ] 6.5%	_____	a) overestimate b) underestimate
5.	[ ] $13.15	3%	_____	a) overestimate b) underestimate
6.	[ ] $65.98	7%	_____	a) overestimate b) underestimate
7.	[ ] $5,898	[ ] 4.75%	_____	a) overestimate b) underestimate
8.	[ ] $684	6%	_____	a) overestimate b) underestimate

Percent Applications

▶ Use your estimation skills to make a reasonable estimate. Then indicate whether your answer is an overestimate or underestimate.

20% Off

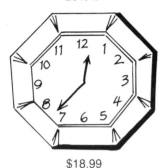

$18.99

1. About how much do you save if you buy the clock at 20% off? _____

   a) overestimate **b)** underestimate

30% Off

$2.19

2. About how much do you save if you buy the work gloves at 30% off?

_____

   a) overestimate **b)** underestimate

9.75% Off

$379

3. About how much do you save if you buy the bracelet at 9.75% off? _____

   a) overestimate **b)** underestimate

4. About how much is a 15% tip on a dinner bill of $43.94? _____

   a) overestimate **b)** underestimate

$275.75

5. The sales tax is 5.75% of your purchase of $275.75. About how much is the sales tax? _____

   a) overestimate **b)** underestimate

40% Off

$9.88

6. About how much will you save if you buy the hose at 40% off? _____

   a) overestimate **b)** underestimate

Using 50% and 25%

Fractional equivalents for 50% and 25% are sometimes useful when working with percents.

$$50\% = \frac{1}{2}$$

Example A

Find 50% of 8.
Think: $\frac{1}{2} \times 8 = 4$

Multiplying by $\frac{1}{2}$ is the same as dividing by 2.

$$25\% = \frac{1}{4}$$

Example B

Find 25% of 40.
Think: $\frac{1}{4} \times 40 = 10$

Multiplying by $\frac{1}{4}$ is the same as dividing by 4.

▶ Change each percent to a fraction and multiply. Work all problems mentally.

$\boxed{\frac{1}{2}}$

1. 50% of 10 = _____

$\boxed{\frac{1}{4}}$

2. 25% of 16 = _____

3. 25% of 80 = _____

4. 50% of 60 = _____

5. 50% of 24 = _____

6. 25% of 360 = _____

7. 50% of 180 = _____

8. 25% of 840 = _____

9. 50% of 660 = _____

10. 25% of 200 = _____

Percent Equivalents

Remembering the fractional equivalent for $33\frac{1}{3}\%$ will sometimes be useful when working with percents.

Example

Find $33\frac{1}{3}\%$ of 12.
Think: $\frac{1}{3} \times 12 = 4$

Multiplying by $\frac{1}{3}$ is the same as dividing by 3.

$$33\frac{1}{3}\% = \frac{1}{3}$$

▶ Change $33\frac{1}{3}\%$ to a fraction and multiply. Work all problems mentally.

$\boxed{\frac{1}{3}}$

1. $33\frac{1}{3}\%$ of $12 = $ _____

2. $33\frac{1}{3}\%$ of $90 = $ _____

3. $33\frac{1}{3}\%$ of $180 = $ _____

4. $33\frac{1}{3}\%$ of $600 = $ _____

$$66\frac{2}{3}\% = \frac{2}{3}$$

$\boxed{\frac{2}{3}}$

Find $66\frac{2}{3}\%$ of 24.
Think: $\frac{1}{3} \times 24 = 8$,
so $\underbrace{\frac{2}{3} \times 24 = 16}_{2 \times 8}$

$$75\% = \frac{3}{4}$$

$\boxed{\frac{3}{4}}$

Find 75% of 36.
Think: $\frac{1}{4} \times 36 = 9$,
so $\underbrace{\frac{3}{4} \times 36 = 27}_{3 \times 9}$

▶ Change $66\frac{2}{3}\%$ or 75% to the fractional equivalent and multiply. Work all problems mentally.

$\boxed{\frac{2}{3}}$

5. $66\frac{2}{3}\%$ of $18 = $ _____

6. $66\frac{2}{3}\%$ of $120 = $ _____

$\boxed{\frac{3}{4}}$

7. 75% of $24 = $ _____

8. 75% of $800 = $ _____

Percents and Compatible Numbers

$$25\% = \tfrac{1}{4} \qquad\qquad 50\% = \tfrac{1}{2} \qquad\qquad 33\tfrac{1}{3}\% = \tfrac{1}{3}$$

Compatible numbers are numbers that make it easy to figure "in your head." The above fractional equivalents for the common percents can often be used with compatible numbers to produce good estimates.

▶ Choose a number that is compatible with the equivalent fraction in each problem.

1. Find $33\tfrac{1}{3}\%$ of 17. [18]

(18 makes it easy to multiply by $\tfrac{1}{3}$.)

$\dfrac{\square}{\square} \times \underline{\ 18\ } = \underline{\qquad}_{\text{estimate}}$

2. Find 50% of 13. [12 or 14]

$\dfrac{\square}{\square} \times \underline{\qquad} = \underline{\qquad}_{\text{estimate}}$

3. Find 25% of 21. $\square$

$\dfrac{\square}{\square} \times \underline{\qquad} = \underline{\qquad}_{\text{estimate}}$

4. Find 50% of 19. $\square$

$\dfrac{\square}{\square} \times \underline{\qquad} = \underline{\qquad}_{\text{estimate}}$

5. Find 25% of 39. $\square$

$\dfrac{\square}{\square} \times \underline{\qquad} = \underline{\qquad}_{\text{estimate}}$

6. Find $33\tfrac{1}{3}\%$ of 5. $\square$

$\dfrac{\square}{\square} \times \underline{\qquad} = \underline{\qquad}_{\text{estimate}}$

7. Find 50% of 49. $\square$

$\dfrac{\square}{\square} \times \underline{\qquad} = \underline{\qquad}_{\text{estimate}}$

8. Find 25% of 81. $\square$

$\dfrac{\square}{\square} \times \underline{\qquad} = \underline{\qquad}_{\text{estimate}}$

Use Compatible Numbers

$$66\tfrac{2}{3}\% = \tfrac{2}{3} \qquad\qquad 75\% = \tfrac{3}{4}$$

You can find reasonable estimates quickly by using fractional equivalents of $66\tfrac{2}{3}\%$ and 75% with compatible numbers.

Example A

$\boxed{27}$

Find $66\tfrac{2}{3}\%$ of 26.
Think: $\tfrac{2}{3} \times 27$
If $\tfrac{1}{3} \times 27 = 9$,
then $\underbrace{\tfrac{2}{3} \times 27 = 18}_{2 \times 9}$ estimate

Example B

$\boxed{32}$

Find 75% of 33.
Think: $\tfrac{3}{4} \times 32$
If $\tfrac{1}{4} \times 32 = 8$,
then $\underbrace{\tfrac{3}{4} \times 32 = 24}_{3 \times 8}$ estimate

---

▶ Change each percent to a fraction. Find a compatible number and then multiply the 2 numbers.

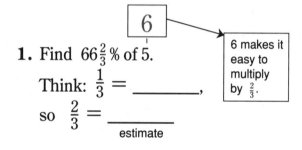

1. Find $66\tfrac{2}{3}\%$ of 5.

$\boxed{6}$ → 6 makes it easy to multiply by $\tfrac{2}{3}$.

Think: $\tfrac{1}{3} = $ _____,

so $\tfrac{2}{3} = $ _____ estimate

2. Find 75% of 19.

Think: $\tfrac{1}{4} = $ _____,

so $\tfrac{3}{4} = $ _____ estimate

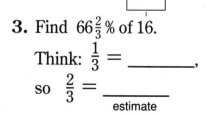

3. Find $66\tfrac{2}{3}\%$ of 16.

Think: $\tfrac{1}{3} = $ _____,

so $\tfrac{2}{3} = $ _____ estimate

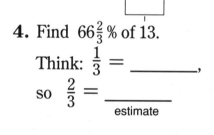

4. Find $66\tfrac{2}{3}\%$ of 13.

Think: $\tfrac{1}{3} = $ _____,

so $\tfrac{2}{3} = $ _____ estimate

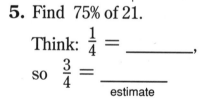

5. Find 75% of 21.

Think: $\tfrac{1}{4} = $ _____,

so $\tfrac{3}{4} = $ _____ estimate

6. Find $66\tfrac{2}{3}\%$ of 17.

Think: $\tfrac{1}{3} = $ _____,

so $\tfrac{2}{3} = $ _____ estimate

Using Easy Percents

Sometimes you will want to use compatible percents to estimate.

$\frac{1}{4}$	$\frac{1}{2}$	$\frac{3}{4}$	$\frac{1}{3}$	$\frac{2}{3}$
25%	50%	75%	$33\frac{1}{3}\%$	$66\frac{2}{3}\%$
26%	**47%**	**73%**	**35%**	**68%**

Example A

25%

Find 24% of 40.

25%

Think: $\frac{1}{4} \times 40 = 10$
estimate

Example B

50%

Find 48% of 120.

50%

Think: $\frac{1}{2} \times 120 = 60$
estimate

Example C

$33\frac{1}{3}\%$

Find 36% of 60.

$33\frac{1}{3}\%$

Think: $\frac{1}{3} \times 60 = 20$
estimate

▶ Use a compatible percent to estimate for each of the following problems.

 50%

1. Find 52% of 200.

$$\frac{\square}{\square} \times 200 = \underline{\hspace{2cm}}$$
estimate

2. Find 27% of 800.

$$\frac{\square}{\square} \times 800 = \underline{\hspace{2cm}}$$
estimate

3. Find 32% of 6,000.

$$\frac{\square}{\square} \times 6{,}000 = \underline{\hspace{2cm}}$$
estimate

4. Find 65% of 30.

$$\frac{\square}{\square} \times \underline{\hspace{1.5cm}} = \underline{\hspace{2cm}}$$
estimate

5. Find 73% of 400.

$$\frac{\square}{\square} \times \underline{\hspace{1.5cm}} = \underline{\hspace{2cm}}$$
estimate

6. Find 26% of 12,000.

$$\frac{\square}{\square} \times \underline{\hspace{1.5cm}} = \underline{\hspace{2cm}}$$
estimate

Estimating with Compatible Numbers

Sometimes you will need to adjust both numbers.

SALE
35% OFF

Regular Price
$6.19

Example

About how much will you save by buying the vegetable flat on sale?

Step 1

Find the fractional equivalent that is close to 35%.

$$33\tfrac{1}{3}\% = \tfrac{1}{3}$$

Find 35% of $6.19.
Change 35% to $\tfrac{1}{3}$.

Step 2

Change $6.19 to a number compatible with $\tfrac{1}{3}$.

| $\tfrac{1}{3}$ | $6 |

Find 35% of $6.19.
$$\tfrac{1}{3} \times \$6 = \$2$$
estimate

You will save about $2.

▶ Find estimates for the following problems using the steps shown above.

1. Find 35% of $12.39.

estimate

2. Find 24% of $7.89.

estimate

3. Find 55% of $79.99.

estimate

4. Find 65% of $59.17.

estimate

5. Find 33% of $15.95.

estimate

6. Find 76% of $25.83.

estimate

Practice Your Skills

▶ Estimate each answer. Then show whether your answer is an overestimate or underestimate.

1. Find 35% of $15.95. _____ **a)** overestimate   **b)** underestimate

estimate

2. Find 25% of $82.35. _____ **a)** overestimate   **b)** underestimate

estimate

3. Find 50% of $489. _____ **a)** overestimate   **b)** underestimate

estimate

4. Find 77% of $320. _____ **a)** overestimate   **b)** underestimate

estimate

5. Find 34% of $9.53. _____ **a)** overestimate   **b)** underestimate

estimate

6. Find 65% of $900. _____ **a)** overestimate   **b)** underestimate

estimate

7. Find 48% of $15.75. _____ **a)** overestimate   **b)** underestimate

estimate

8. Find 26% of $2,000. _____ **a)** overestimate   **b)** underestimate

estimate

Find the Percent

To find what percent one number is of another,

- Write a fraction comparing the part to the total.
- Rename the fraction in hundredths.
- Write it as a percent.

▶ For each problem, write a fraction to compare the part to the total. Then change the fraction to a percent.

Question	**Think**

1. 100 is what percent of 100?
   ↑   ↑
   part   total

$\dfrac{\text{Part}}{\text{Total}} \rightarrow \dfrac{100}{100} = 100\%$

2. 20 is what percent of 100?

$\dfrac{\text{Part}}{\text{Total}}$ $\dfrac{\square}{100} = $ _____%

3. 10 is what percent of 20?

┌ multiply by 5 ┐

$\dfrac{\boxed{10}}{20} = \dfrac{\square}{100} = $ _____%

└ multiply by 5 ┘

4. 5 is what percent of 20?

$\dfrac{\square}{\square} = \dfrac{\square}{100} = $ _____%

5. 4 is what percent of 16?

$\dfrac{\square}{\square} = \dfrac{\square}{\square} = \dfrac{\square}{100} = $ _____%

simplified

6. 8 is what percent of 40?

$\dfrac{\square}{\square} = \dfrac{\square}{\square} = \dfrac{\square}{100} = $ _____%

simplified

7. 6 is what percent of 20?

$\dfrac{\square}{\square} = \dfrac{\square}{100} = $ _____%

Fractions Compatible with Common Percents

$$\frac{1}{4} = 25\% \qquad \frac{1}{2} = 50\% \qquad \frac{3}{4} = 75\%$$

$$\frac{1}{3} = 33\frac{1}{3}\% \qquad \frac{2}{3} = 66\frac{2}{3}\% \qquad \frac{1}{10} = 10\%$$

A compatible fraction is a fraction that makes it easy to figure percents in your head.

▶ Solve each of the following problems by choosing a compatible fraction and an equivalent percent from the above list.

1. $\frac{21}{30}$ about $\frac{20}{30}$ $\frac{2}{3}$ (compatible fraction) $= 66\frac{2}{3}$ % (estimate)

2. $\frac{5}{19}$ about $\frac{5}{20}$ _____ (compatible fraction) $= 25$ % (estimate)

3. $\frac{16}{30}$ _____ (compatible fraction) $=$ _____ % (estimate)

4. $\frac{7}{27}$ _____ (compatible fraction) $=$ _____ % (estimate)

5. $\frac{4}{13}$ _____ (compatible fraction) $=$ _____ % (estimate)

6. $\frac{3}{29}$ about $\frac{3}{30}$ _____ (compatible fraction) $=$ _____ % (estimate)

7. $\frac{8}{25}$ about $\frac{8}{24}$ _____ (compatible fraction) $=$ _____ % (estimate)

8. $\frac{5}{11}$ _____ (compatible fraction) $=$ _____ % (estimate)

9. $\frac{10}{16}$ _____ (compatible fraction) $=$ _____ % (estimate)

10. $\frac{4}{42}$ _____ (compatible fraction) $=$ _____ % (estimate)

Estimate the Percent

To estimate what percent one number is of another,

- Compare the part to the total.
- Think of a compatible fraction.
- Estimate the equivalent percent.

Example

A $2 tip is what percent of a $21 food bill?

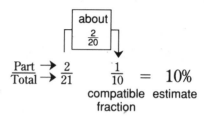

$$\frac{1}{4} = 25\% \qquad \frac{3}{4} = 75\% \qquad \frac{2}{3} = 66\frac{2}{3}\%$$

$$\frac{1}{2} = 50\% \qquad \frac{1}{3} = 33\frac{1}{3}\% \qquad \frac{1}{10} = 10\%$$

▶ Solve the following problems by finding the compatible fraction and estimating the equivalent percent.

1. 5 is what percent of 49?

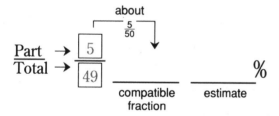

2. 20 is what percent of 31?

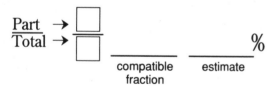

3. 12 is what percent of 25?

4. 4 is what percent of 15?

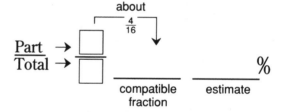

5. 41 is what percent of 62?

6. 31 is what percent of 42?

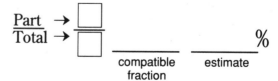

Use Compatible Fractions

▶ Use compatible fractions to estimate each percent. Work all problems mentally.

Marked Down $4

$17

1. About what percent of the cost has been marked off? _____

2. Dawn missed 3 out of 29 questions. About what percent of the problems did she miss? _____

3. Jim had 11 hits out of 23 times at bat. About what percent of his times at bat were hits? _____

4. The Bulldogs lost 7 out of 18 football games. About what percent of the games did the team lose? _____

5. Sarah made 9 out of 13 free throws. About what percent of her attempted free throws did she make? _____

Was $140
Now $69

6. About what percent of the cost has been marked off? _____

7. 9 out of 35 apples were spoiled. About what percent of the apples were spoiled? _____

(Read this problem carefully!)

8. The soccer team lost only 1 game out of 11. About what percent of the games did it win? _____

Find the Total, or 100%

To find the total, you must find 100%. Sometimes you can multiply the percent by a number to get 100%. This will help you to estimate the total.

Sale
20% Off
Regular Price

Example

Kimberly bought a bracelet on sale and saved $6.00. What was the regular price of the bracelet?

Find the total,
or 100%

20% of _____ is $6
percent   part

Think: 20% = $6
        100% = $30

5 × 20%   5 × $6

The regular price of the bracelet was $30.

▶ Find the total (100%) for each of the following problems. Work all problems mentally.

1. 10% of ___?___ is 4
Think: $10\% = 4$,
so $100\% =$ _____

10 × 10%   10 × 4

5. 5% of ___?___ is 8
Think: $5\% = 8$,
so $100\% =$ _____

2. 20% of ___?___ is 7
Think: $20\% =$ _____,
so $100\% =$ _____

6. $33\frac{1}{3}$% of ___?___ is 10
Think: $33\frac{1}{3}\% =$ _____,
so $100\% =$ _____

3. 25% of ___?___ is 40
Think: $25\% =$ _____,
so $100\% =$ _____

7. 10% of ___?___ is 20
Think: $10\% =$ _____,
so $100\% =$ _____

4. 50% of ___?___ is 90
Think: $50\% =$ _____,
so $100\% =$ _____

8. 25% of ___?___ is 500
Think: $25\% =$ _____,
so $100\% =$ _____

Mentally Find the Total

Sometimes when estimating the total, you need to work with the percent before you can change it to 100%.

Example A

total, or 100%

80% of _____?_____ is 16

Think:

Divide by 8. → 80% = 16 ← Divide by 8.
10% = 2
100% = 20

Example B

total, or 100%

75% of _____?_____ is 12

Think:

Divide by 3. → 75% = 12 ← Divide by 3.
25% = 4
100% = 16

▶ Solve each of the following problems by looking for a pattern to find the total (100%). Work all problems mentally.

1. 60% of _____?_____ is 15

Divide by 3. → 60% = 15 ← Divide by 3.
20% = _____
100% = _____

5. $66\frac{2}{3}$% of _____?_____ is 10

Divide by 2. → $66\frac{2}{3}$% = 10 ← Divide by 2.
$33\frac{1}{3}$% = _____
100% = _____

2. 40% of _____?_____ is 80

10% = _____
100% = _____

6. 30% of _____?_____ is 27

10% = _____
100% = _____

3. 70% of _____?_____ is 49

10% = _____
100% = _____

7. 4% of _____?_____ is 20

1% = _____
100% = _____

4. 90% of _____?_____ is 27

10% = _____
100% = _____

8. 75% of _____?_____ is 24

25% = _____
100% = _____

Estimate to Find the Total

▶ For each of the following problems, use compatible numbers to estimate the total. Work all problems mentally.

1. 50% of ___?___ is $31.05 $30

100% = _____
       estimate

2. 10% of ___?___ is $5.10

100% = _____
       estimate

3. 40% of ___?___ is $7.89

10% = _____
    estimate

100% = _____
      estimate

4. 25% of ___?___ is $98.79

100% = _____
      estimate

5. 75% of ___?___ is $23.95

25% = _____
    estimate

100% = _____
      estimate

6. 20% of ___?___ is $9.98 $10

100% = _____
       estimate

7. 5% of ___?___ is $2.95

100% = _____
       estimate

8. 3% of ___?___ is $12.35

1% = _____
   estimate

100% = _____
      estimate

9. 10% of ___?___ is $49.50

100% = _____
      estimate

10. 4.8% of ___?___ is $19.75   5%

5% = _____
   estimate

100% = _____
      estimate

10 APPLICATIONS

Estimate the Discounts

A discount is how much something has been reduced in price.

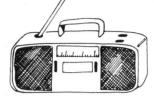

25% Off

To estimate the discount,
• Change the percent to a fraction.
• Find a compatible dollar amount.
• Multiply the 2 numbers.

If this radio regularly sells
for $78.99, about how much will you save?

$\frac{1}{4}$ $80

25% of $78.99 = Discount

$\frac{1}{4} \times \$80 = \$20$
estimate

The discount on the radio is about $20.

▶ Estimate the discount for each of the following items. Then indicate whether your answer is an overestimate or an underestimate.

	Item	Regular Price	Percent Off	Discount (Estimate)	Circle the Answer
1.	Camcorder	$985.97	20%	_____	a) overestimate b) underestimate
2.	Earrings	$23.10	25%	_____	a) overestimate b) underestimate
3.	Sunglasses	$8.75	$33\frac{1}{3}$%	_____	a) overestimate b) underestimate
4.	Jeans	$16.49	50%	_____	a) overestimate b) underestimate
5.	12-HP Tractor	$1,299	25%	_____	a) overestimate b) underestimate
6.	Lawnmower	$375.99	10%	_____	a) overestimate b) underestimate

Estimate the Best Buy

▶ Sometimes you may want to compare prices to get the best value for your money. Use your estimation skills to answer the following problems.

1. Does Food Value or Apple Foods have the better buy? _____

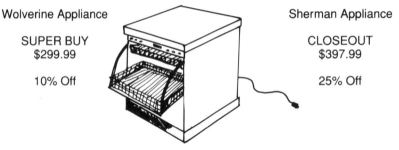

Food Value

Regular $3.99
Save 5%

Apple Foods

Regular $4.50
Save 10%

Save about: _____   Save about: _____
Estimated price: _____   Estimated price: _____

2. Does Wolverine or Sherman have the better buy? _____

Wolverine Appliance

SUPER BUY
$299.99

10% Off

Sherman Appliance

CLOSEOUT
$397.99

25% Off

Save about: _____   Save about: _____
Estimated price: _____   Estimated price: _____

3. Does Nelson or Fred's have the better buy? _____

Nelson Hardware

50% Off

Sale

$28.95

Fred's Hardware

33 $\frac{1}{3}$ % Off

Sale

$26.95

Save about: _____   Save about: _____
Estimated price: _____   Estimated price: _____

Do You Have Enough Money?

▶ Use your estimation skills to answer the following questions.

30% Off
Marked Price

$17.95

1. Will $10 be enough to buy the clock?

_____

Sale
25% Off
Regular Price

$28.35

2. Will $20 be enough to buy the watch?

_____

3. If the sales tax is 6%, will $5 be enough? _____

4. If Joy pays a 15% tip, will $50.00 be enough to pay for the dinner bill?

_____

$169.88

5. Will $80.00 be enough for a 50% down payment? _____

Regular Price: $2.45
25% Off

6. Will the sale price for the coffee mug be under $2? _____

Review of Percent Estimation

▶ Answer each of the following questions by estimating. Be flexible. There are many ways to estimate. Before you start a problem, take a minute to think about the numbers involved. Use a method that is quick and easy to do in your head.

20% Off
$28.93

1. About how much will you save by buying the fishing reel on sale?

_____

Total: $42.95

2. Ramona paid a 15% tip. About how much did she add for the tip?

_____

3. José bought a battery for $47.19 and paid a 6% sales tax. About how much was the sales tax? _____

4. Reggie's monthly income is $1,973. He wants to budget 25% of his income for food. About how much money will he budget for food each month?

_____

Sale
10% Off

5. Tom bought a gas grill on sale and saved $6.95. About what was the regular price of the grill? _____

Sale
35% Off
$58.70

6. About how much will you save by buying the baseball glove on sale?

_____

7. A necklace was priced $38.98. It was then marked down $10.00. About what percent was marked off?

8. Anita sold $58.90 worth of magazines and was paid a 50% commission. About how much money was she paid in commissions? _____

ANSWER KEY

Page 1: Learning About Estimation

1. b) 100 **2. c)** $\frac{1}{4}$ **3. a)** $30

Page 2: Decide When to Estimate

1. a) exact **4. b)** estimate
2. b) estimate **5. a)** exact
3. b) estimate **6. b)** estimate

Page 3: Shade the Fractions

1. $\frac{3}{4}$

2. $\frac{2}{3}$

3. $\frac{7}{8}$

4. $\frac{1}{5}$

5. $\frac{6}{10}$

6. $\frac{3}{6}$

Page 4: Estimate the Fractions

1. $\frac{4}{5}$ **5.** $\frac{9}{10}$

2. $\frac{1}{2}$ **6.** $\frac{1}{3}$

3. $\frac{2}{3}$ **7.** $\frac{3}{4}$

4. $\frac{1}{6}$ **8.** $\frac{1}{2}$

Page 5: Fractions Close to 0, $\frac{1}{2}$, or 1

1. 1 **3.** 0 **5.** $\frac{1}{2}$
2. $\frac{1}{2}$ **4.** 1 **6.** 0

Page 6: Circle the Fractions

1. $\frac{2}{20}$; $\frac{1}{9}$; $\frac{2}{30}$; $\frac{4}{100}$; $\frac{1}{16}$
2. $\frac{6}{13}$; $\frac{5}{9}$; $\frac{9}{16}$; $\frac{4}{7}$
3. $\frac{4}{5}$; $\frac{21}{23}$; $\frac{9}{10}$; $\frac{15}{16}$; $\frac{8}{9}$

Page 7: Working with $\frac{1}{2}$

1. < **4.** > **7.** <
2. < **5.** > **8.** >
3. > **6.** <

Page 8: Greater than or Less than $\frac{1}{2}$

A. less; twice 5 is less than 12.
B. more; twice 9 is more than 16.

1. b) less than $\frac{1}{2}$ **4. a)** greater than $\frac{1}{2}$
2. b) less than $\frac{1}{2}$ **5. a)** greater than $\frac{1}{2}$
3. a) greater than $\frac{1}{2}$ **6. b)** less than $\frac{1}{2}$

Page 9: Use Your Skills

1. a) 10; **b)** 4; **c)** 5; **d)** 14; **e)** 6
2. a) 4 or 5; **b)** 2 or 3; **c)** 6 or 7; **d)** 3 or 4; **e)** 5 or 6
3. a) 6, 7, 8, 9, or 10; **b)** 9, 8, 7, or 6; **c)** 7, 6, or 5;
d) 5, 6, or 7; **e)** 3 or 4
4. a) 1; **b)** 5, 6, 7 . . . ; **c)** 11, 12, 13 . . . ; **d)** 1, 2, or 3;
e) 17, 18, 19 . . .

Page 10: Rounding Mixed Numbers

1. 1 **4.** 5 **7.** 6
2. 10 **5.** 2 **8.** 4
3. 7 **6.** 6

Page 11: Rounding to Estimate the Sum

1.
$3\frac{4}{5}$ rounds to 4
$+ 5\frac{1}{6}$ rounds to $+ 5$
Estimate: 9

4.
$6\frac{1}{2}$ rounds to 7
$4\frac{5}{6}$ rounds to 5
$+ 8\frac{1}{5}$ rounds to $+ 8$
Estimate: 20

2.
$6\frac{1}{4}$ rounds to 6
$+ 3\frac{3}{5}$ rounds to $+ 4$
Estimate: 10

5.
$9\frac{2}{3}$ rounds to 10
$3\frac{3}{8}$ rounds to 3
$+ 2\frac{5}{7}$ rounds to $+ 3$
Estimate: 16

3.
$7\frac{3}{10}$ rounds to 7
$+ 2\frac{2}{3}$ rounds to $+ 3$
Estimate: 10

6.
$3\frac{5}{8}$ rounds to 4
$8\frac{1}{2}$ rounds to 9
$+ 1\frac{3}{4}$ rounds to $+ 2$
Estimate: 15

Page 12: Grouping Fractions to Estimate

1. Estimate: $2\frac{1}{2}$ **4.** Estimate: $3\frac{1}{2}$
2. Estimate: 2 **5.** Estimate: $1\frac{1}{2}$–2
3. Estimate: 2 **6.** Estimate: 2–$2\frac{1}{2}$

Page 13: Adjusting Front-End Estimation

1. Front-end estimate: 12
   Adjusted estimate: 13
2. Front-end estimate: 15
   Adjusted estimate: 17
3. Front-end estimate: 15
   Adjusted estimate: 16
4. Front-end estimate: 16
   Adjusted estimate: 18
5. Front-end estimate: 16
   Adjusted estimate: 18
6. Front-end estimate: 11
   Adjusted estimate: 13

Page 14: More Adjusting Front-End Estimation

1. Front-end estimate: 13
   Adjusted estimate: $14\frac{1}{2}$
2. Front-end estimate: 14
   Adjusted estimate: 16
3. Front-end estimate: 18
   Adjusted estimate: 19
4. Front-end estimate: 11
   Adjusted estimate: $13\frac{1}{2}$
5. Front-end estimate: 10
   Adjusted estimate: 11
6. Front-end estimate: 18
   Adjusted estimate: 20
7. Front-end estimate: 11
   Adjusted estimate: $12\frac{1}{2}$
8. Front-end estimate: 6
   Adjusted estimate: $8\frac{1}{2}$

Page 15: Estimating Larger Mixed Numbers

Estimates will vary.

1. 60–70
2. 40–45
3. 70–80
4. 110–120
5. 1,000–1,125
6. 1,100–1,200
7. 800–900
8. 1,000–1,050

Page 16: Practice Your Skills

1. a) Front-end: 16
   b) Adjusted: $18\frac{1}{2}$
   c) Rounded: 18

2. a) Front-end: 9
   b) Adjusted: 10
   c) Rounded: 10

3. a) Front-end: 12
   b) Adjusted: $14\frac{1}{2}$
   c) Rounded: 15

4. a) Front-end: 11
   b) Adjusted: 14
   c) Rounded: 14

5. a) Front-end: 18
   b) Adjusted: 20
   c) Rounded: 19

6. a) Front-end: 25
   b) Adjusted: 27
   c) Rounded: 27

Page 17: Is the Answer Reasonable?

1. b) 6 hours
2. c) 4 pounds
3. b) 3 inches
4. a) 16 gallons

Page 18: Rounding Mixed Numbers to Subtract

1. $6\frac{2}{3}$ rounds to   7
   $-1\frac{1}{8}$ rounds to -1
   Estimate: 6

2. $4\frac{3}{5}$ rounds to   5
   $-2\frac{7}{8}$ rounds to -3
   Estimate: 2

5. $7\frac{5}{8} - 5\frac{5}{12}$ $\boxed{8}$ $\boxed{5}$
   Estimate: 3

6. $6\frac{5}{7} - 2\frac{3}{5}$ $\boxed{7}$ $\boxed{3}$
   Estimate: 4

3. Estimate: 3
4. Estimate: 1
7. Estimate: 8
8. Estimate: 5

Page 19: Adjust the Difference

1. b) less than 3
2. a) greater than 5
3. b) less than 2
4. b) less than 7
5. a) greater than 3
6. a) greater than 1

Page 20: Subtracting Larger Mixed Numbers

Estimates will vary.

1. 25–30
2. 40–50
3. 75–80
4. 25–30
5. 80–85
6. 170–190
7. 80–90
8. 610–700

Page 21: Estimate to Add or Subtract

Estimates will vary.

1. 4 hours
2. a) 6 miles
   b) 9 miles
   c) 6 miles
3. 12 feet
4. a) 11 pounds
   b) 7 pounds
   c) 17 or 18 pounds

Page 22: Real-World Applications

Remember: any estimate that is close is acceptable.

1. 9 yards
2. 500 miles
3. 12 inches
4. 15 gallons
5. 3 pounds
6. 17 pounds

Page 23: Does the Answer Make Sense?

1. b) $60
2. b) 13 gallons
3. b) 600 students
4. b) $3.50

Page 24: Rounding Mixed Numbers to Multiply

1. $8\frac{2}{3} \times 4\frac{7}{8}$
$9 \times 5 = 45$

2. $4\frac{3}{7} \times 6\frac{3}{5}$
$4 \times 7 = 28$

3. $1\frac{1}{8} \times 2\frac{2}{9}$
$1 \times 2 = 2$

4. $3\frac{2}{3} \times 7\frac{1}{4}$
$4 \times 7 = 28$

5. $9\frac{1}{6} \times 4\frac{5}{8}$
$9 \times 5 = 45$

6. $4\frac{7}{9} \times 7\frac{1}{2}$
$5 \times 8 = 40$

7. $8\frac{3}{4} \times 2\frac{1}{6}$
$9 \times 2 = 18$

8. $5\frac{2}{5} \times 6\frac{4}{9}$
$5 \times 6 = 30$

Page 25: Round to Multiply

1. $5 \times 3\frac{2}{5}$
$5 \times 3 = 15$

2. $3\frac{7}{10} \times 10$
$4 \times 10 = 40$

3. $2\frac{2}{3} \times 4$
$3 \times 4 = 12$

4. $8 \times 4\frac{1}{8}$
$8 \times 4 = 32$

5. $7\frac{3}{5} \times 6$
$8 \times 6 = 48$

6. $3 \times 8\frac{1}{2}$
$3 \times 9 = 27$

Page 26: Rounding for Sensible Answers

1. $2\frac{1}{4} \times \$3.89$
$2 \times \$4 = \$8$

2. $5\frac{3}{4} \times \$.99$
$6 \times \$1 = \$6$

3. $6\frac{3}{5} \times \$4.99$
$7 \times \$5 = \$35$

4. $1\frac{5}{8} \times \$5.78$
$2 \times \$6 = \$12$

Page 27: Estimating with $\frac{1}{2}$

Estimates will vary. Exact amounts are given in parentheses.

1. a) 5 **b)** 5
2. a) 3 **b)** 3
3. a) 10 **b)** 10
4. a) 25 **b)** 25
5. $5.00 ($4.63)
6. $9.00–$10.00 ($9.45)
7. $.50 ($.47)
8. $10.00 ($9.99)
9. $3.00 ($3.20)
10. $4.00 ($3.84)
11. $4.00 ($4.13)
12. $7.00 ($6.95)

Page 28: Use Compatible Numbers

Estimates will vary. Exact amounts are given in parentheses.

1. $3.00 ($2.99)
2. $7.00 ($6.86)
3. $3.00 ($2.83)
4. $.20 ($.20)
5. $9.00 ($9.38)
6. $4.00 ($3.76)
7. $10.00 ($10.36)
8. $9.00 ($9.40)
9. $50.00 ($49.45)
10. $10.00 ($10.45)

Page 29: Estimate What You Save

Estimates will vary. Exact amounts are given in parentheses.

1. $10.00 ($9.93)
2. $8.00 ($8.37)
3. $5.00 ($5.16)
4. $7.00 ($6.79)
5. $50.00 ($49.45)
6. $10.00 ($10.00)

Page 30: Estimate What You Would Pay

Estimates will vary. Exact amounts are given in parentheses.

	Save	Pay
1.	$1.00 ($.97)	$3.00 ($2.92)
2.	$25.00 ($24.48)	$25.00 ($24.48)
3.	$9.00 ($8.93)	$18.00 ($17.86)
4.	$20.00 ($19.71)	$80.00 ($78.84)
5.	$6.00 ($6.00)	$18.00 ($17.98)

Page 31: Figuring Costs

Estimates will vary. Exact amounts are given in parentheses.

1. If you save $\frac{1}{4}$, then you must pay $\frac{3}{4}$.
$\frac{3}{4} \times \$40 = \$30$
You would pay about $30 for the radio. ($31.26)

2. If you save $\frac{1}{3}$, then you must pay $\frac{2}{3}$.
$\frac{2}{3} \times \$90 = \$60$
You would pay about $60 for the camera. ($59.33)

3. If you save $\frac{1}{2}$, then you must pay $\frac{1}{2}$.
$\frac{1}{2} \times \$18 = \$9$
You would pay about $9 for the calculator. ($8.99)

4. If you save $\frac{1}{5}$, then you must pay $\frac{4}{5}$.
$\frac{4}{5} \times \$30 = \$24$
You would pay about $24 for the basketball. ($23.01)

Page 32: Clustering

1. a) 10 miles
b) 50 miles

2. a) 5 pounds
b) 30 pounds

Page 33: Estimating to Divide

1. b) 8
2. c) 3
3. a) 9
4. a) 2

Page 34: Rounding to Estimate

1. $9 \div 2 = \frac{9}{2} = 4\frac{1}{2}$
2. $5 \div 7 = \frac{5}{7}$
3. $9 \div 3 = 3$
4. $2 \div 2 = 1$
5. $15 \div 4 = \frac{15}{4} = 3\frac{3}{4}$
6. $10 \div 2 = 5$
7. $3 \div 5 = \frac{3}{5}$
8. $4 \div 3 = \frac{4}{3} = 1\frac{1}{3}$

Page 35: Real-Life Applications

1. About $1\frac{1}{2}$ cups of sugar
2. Two $2\frac{3}{4}$-foot strips
3. About 8 miles
4. About $2\frac{1}{2}$ yards
5. About 4 rows

Page 36: Estimate the Costs

Estimates will vary. Exact amounts are given in parentheses.

1. a) $.47   b) $3.50 ($3.47)
2. a) $1.19   b) $10.00 ($11.31)
3. a) $5.39   b) $15.00 ($17.52)
4. a) $7.84   b) $24.00 ($21.56)
5. a) $.89   b) $2.00 ($2.00)
6. a) $3.95   b) $16.00 ($14.81)
7. a) $6.13   b) $30.00 ($28.35)
8. a) $4.78   b) $5.00 ($5.38)

Page 37: Is the Answer Reasonable?

Estimates will vary. Exact amounts are given in parentheses.

1. 14 gallons ($14\frac{3}{8}$)
2. 5 inches ($5\frac{1}{4}$)
3. 28 pounds ($27\frac{1}{2}$)
4. a) $6.00–$7.00 ($6.49)
   b) $18.00–$21.00 ($19.46)
5. 2 cups ($2\frac{1}{12}$)
6. 8 pounds ($7\frac{3}{8}$)

Page 38: Review of Fraction Estimation

Estimates will vary. Exact amounts are given in parentheses.

1. 180 miles ($194\frac{1}{4}$)
2. 7 inches ($7\frac{1}{8}$)
3. 34 miles ($33\frac{7}{8}$)
4. 4 miles ($4\frac{5}{8}$)
5. $15.00 ($14.98)
6. $26–$28 ($26.08)
7. 15 gallons ($14\frac{3}{4}$)
8. 2 cups (1.93)

Page 39: Estimating with Percents

1. b) $1,000
2. d)
3. a) 25%
4. b) $1.50

Page 40: Meaning of Percent

1. 30
2. 100
3. 1%
4. 75
5. 33%
6. $.75 = \frac{75}{100}$
7. $\frac{25}{100} = 25\%$
8. $\frac{50}{100} = 50\%$

9. 40%
10. 91%

Page 40: Meaning of Percent (continued)

11. 35%
12. 59%

Page 41: Percents Close to 100%, 10%, or 1%

1. 100% of 400 = 400
   10% of 400 = 40
   1% of 400 = 4
2. 100% of 500 = 500
   10% of 500 = 50
   1% of 500 = 5
3. 100% of 150 = 150
   10% of 150 = 15
   1% of 150 = 1.5
4. 100% of 86 = 86
   10% of 86 = 8.6
   1% of 86 = .86

Page 42: Finding 100%, 10%, and 1% of a Number

1. c) 72
2. a) .9
3. b) .15
4. b) 93.5
5. c) 398
6. a) .54
7. 1% of 200 = 2
8. 10% of 75 = 7.5
9. 100% of 25 = 25
10. 10% of 17 = 1.7
11. 100% of 36 = 36
12. 1% of 400 = 4

Page 43: Using 10%

1. $12
2. Find 60% of 90.
   10% = 9
   60% = 54
3. Find 90% of $20.
   10% = $2
   90% = $18
4. Find 80% of $50.
   10% = $5
   80% = $40
5. $45
6. Find 20% of 300.
   10% = 30
   20% = 60
7. Find 70% of $80.
   10% = $8
   70% = $56
8. Find 20% of 350.
   10% = 35
   20% = 70

Page 44: Using 10% to Estimate

1. 30 Find 80% of $32.65.
   10% = $3
   80% = $24
2. 300 Find 20% of $317.45.
   10% = $30
   20% = $60
3. 50 Find 70% of $47.89.
   10% = $5
   70% = $35
4. 600 Find 40% of $583.00.
   10% = $60
   40% = $240
5. 70 Find 90% of $74.10.
   10% = $7
   90% = $63
6. 200 Find 60% of $179.45.
   10% = $20
   60% = $120

Page 45: Using 1%

1. $12

5. $1.25

2. Find 4% of 200.
1% = $2
4% = $8

6. Find 2% of $12.
1% = $.12
2% = $.24

3. Find 7% of $500.
1% = $5
7% = $35

7. Find 6% of $20.
1% = $.20
6% = $1.20

4. Find 8% of $7,000.
1% = $70
8% = $560

8. Find 3% of $1,500.
1% = $15
3% = $45

Page 46: Using 1% to Estimate

$10

1. Find 5% of $9.89.
1% × $10 = $.10
5% × $10 = $.50

$7

4. Find 3% of $7.25.
1% = $.07
3% = $.21

$70

2. Find 2% of $72.99.
1% = $.70
2% = $1.40

$8,000

5. Find 7% of $8,275.
1% = $80
7% = $560

$5,000

3. Find 8% of $4,750.
1% = $50
8% = $400

$2,000

6. Find 9% of $1,895.
1% = $20
9% = $180

Page 47: Round to Lead Digits

2% | $2,000

1. Find 2.3% of $2,275.
1% = $20 so 2% = $40
b) underestimate

20% | $500

4. Find 17% of $495.
10% = 50 so 20% = $100
a) overestimate

90% | $30

2. Find 87% of $29.15.
10% = $3 so 90% = $27
a) overestimate

2% | $10,000

5. Find 1.75% of $9,750.
1% = $100 so 2% = $200
a) overestimate

10% | $7,000

3. Find 11.75% of $7,299.
10% = $700
b) underestimate

5% | $50

6. Find 5.15% of $54.
1% = $.50 so 5% = $2.50
b) underestimate

Page 48: Estimate a 15% Tip

$40

1. Dinner bill $38.95
10% = $4
+ 5% = + $2
15% = $6

$80

4. Dinner bill $76.12
10% = $8
+ 5% = + $4
15% = $12

$60

2. Dinner bill $62.15
10% = $6
+ 5% = + $3
15% = $9

$5

5. Dinner bill $4.73
10% = $.50
+ 5% = + $.25
15% = $.75

$10

3. Dinner bill $9.53
10% = $1.00
+ 5% = + $.50
15% = $1.50

$100

6. Dinner bill $97.25
10% = $10
+ 5% = + $5
15% = $15

Page 49: Estimate the Sales Tax

Amount of Purchase	Sales Tax Rate	Sales Tax Estimate	Answer
$50			
1. $51.50	6%	$3	**b)** underestimate
$900			
2. $893	4%	$36	**a)** overestimate
$20	6%		
3. $18.32	5.8%	$1.20	**a)** overestimate
$200	7%		
4. $193	6.5%	$14	**a)** overestimate
$10			
5. $13.15	3%	$.30	**b)** underestimate
$70			
6. $65.98	7%	$4.90	**a)** overestimate
$6,000	5%		
7. $5,898	4.75%	$300	**a)** overestimate
$700			
8. $684	6%	$42	**a)** overestimate

Page 50: Percent Applications

Estimates will vary. Exact amounts are given in parentheses.

1. About $4.00 ($3.80)
a) overestimate

3. About $40.00 ($36.95)
a) overestimate

2. About $.60 ($.66)
b) underestimate

4. About $6.00 ($6.59)
b) underestimate

Page 50: Percent Applications (continued)

5. About $18.00 ($15.86)   **6.** About $4.00 ($3.95)
   a) overestimate   **a)** overestimate

Page 51: Using 50% and 25%

1. 5   **5.** 12   **8.** 210
2. 4   **6.** 90   **9.** 330
3. 20   **7.** 90   **10.** 50
4. 30

Page 52: Percent Equivalents

1. 4   **4.** 200   **7.** 18
2. 30   **5.** 12   **8.** 600
3. 60   **6.** 80

Page 53: Percents and Compatible Numbers

$\boxed{18}$ $\boxed{40}$

1. Find $33\frac{1}{3}\%$ of 17.
$\frac{1}{3} \times 18 = 6$
$\boxed{12 \text{ or } 14}$

5. Find 25% of 39.
$\frac{1}{4} \times 40 = 10$
$\boxed{6}$

2. Find 50% of 13.
$\frac{1}{2} \times 12 = 6$ or
$\frac{1}{2} \times 14 = 7$
$\boxed{20}$

6. Find $33\frac{1}{3}\%$ of 5.
$\frac{1}{3} \times 6 = 2$
$\boxed{50}$

3. Find 25% of 21.
$\frac{1}{4} \times 20 = 5$
$\boxed{18 \text{ or } 20}$

7. Find 50% of 49.
$\frac{1}{2} \times 50 = 25$
$\boxed{80}$

4. Find 50% of 19.
$\frac{1}{2} \times 20 = 10$ or
$\frac{1}{2} \times 18 = 9$

8. Find 25% of 81.
$\frac{1}{4} \times 80 = 20$

Page 54: Use Compatible Numbers

$\boxed{6}$ $\boxed{12}$

1. Find $66\frac{2}{3}\%$ of 5.
$\frac{1}{3} = 2$
$\frac{2}{3} = 4$
$\boxed{20}$

4. Find $66\frac{2}{3}\%$ of 13.
$\frac{1}{3} = 4$
$\frac{2}{3} = 8$
$\boxed{20}$

2. Find 75% of 19.
$\frac{1}{4} = 5$
$\frac{3}{4} = 15$
$\boxed{15}$

5. Find 75% of 21.
$\frac{1}{4} = 5$
$\frac{3}{4} = 15$
$\boxed{18}$

3. Find $66\frac{2}{3}\%$ of 16.
$\frac{1}{3} = 5$
$\frac{2}{3} = 10$

6. Find $66\frac{2}{3}\%$ of 17.
$\frac{1}{3} = 6$
$\frac{2}{3} = 12$

Page 55: Using Easy Percents

$\boxed{50\%}$ $\boxed{66\frac{2}{3}\%}$

1. Find 52% of 200.
$\frac{1}{2} \times 200 = 100$
$\boxed{25\%}$

4. Find 65% of 30.
$\frac{2}{3} \times 30 = 20$
$\boxed{75\%}$

2. Find 27% of 800.
$\frac{1}{4} \times 800 = 200$
$\boxed{33\frac{1}{3}\%}$

5. Find 73% of 400.
$\frac{3}{4} \times 400 = 300$
$\boxed{25\%}$

3. Find 32% of 6,000.
$\frac{1}{3} \times 6,000 = 2,000$

6. Find 26% of 12,000.
$\frac{1}{4} \times 12,000 = 3,000$

Page 56: Estimating with Compatible Numbers

Estimates will vary. Exact amounts are given in parentheses.

$\boxed{33\frac{1}{3}\%}$ $\boxed{\$12}$

1. 35% of $12.39
Estimate: $4.00 ($4.34)

$\boxed{25\%}$ $\boxed{\$8}$

2. 24% of $7.89
Estimate: $2.00 ($1.89)

$\boxed{50\%}$ $\boxed{\$80}$

3. 55% of $79.99
Estimate: $40.00 ($43.99)

$\boxed{66\frac{2}{3}\%}$ $\boxed{\$60}$

4. 65% of $59.17
Estimate: $40.00 ($38.46)

$\boxed{33\frac{1}{3}\%}$ $\boxed{\$15}$

5. 33% of $15.95
Estimate: $5.00 ($5.26)

$\boxed{75\%}$ $\boxed{\$24}$

6. 76% of $25.83
Estimate: $18.00 ($19.63)

Page 57: Practice Your Skills

Estimates will vary. Exact amounts are given in parentheses.

$\boxed{33\frac{1}{3}\%}$ $\boxed{\$15}$

1. Find 35% of $15.95.
Estimate: $5.00 ($5.58)
b) underestimate

$\boxed{\$80}$

2. Find 25% of $82.35.
Estimate: $20.00 ($20.59)
b) underestimate

Page 57: Practice Your Skills (continued)

$\boxed{\$500}$

3. Find 50% of $489.
Estimate: $250 ($244.50)
  a) overestimate

$\boxed{75\%}$

4. Find 77% of $320.
Estimate: $240 ($246.40)
  b) underestimate

$\boxed{33\frac{1}{3}\%}$ $\boxed{\$9}$

5. Find 34% of $9.53.
Estimate: $3.00 ($3.24)
  b) underestimate

$\boxed{66\frac{2}{3}\%}$

6. Find 65% of $900.
Estimate: $600 ($585)
  a) overestimate

$\boxed{50\%}$ $\boxed{\$16}$

7. Find 48% of $15.75.
Estimate: $8.00 ($7.56)
  a) overestimate

$\boxed{25\%}$

8. Find 26% of $2,000.
Estimate: $500 ($520)
  b) underestimate

Page 58: Find the Percent

1. $\frac{100}{100} = 100\%$

2. $\frac{20}{100} = 20\%$

3. $\frac{10}{20} = \frac{50}{100} = 50\%$

4. $\frac{5}{20} = \frac{25}{100} = 25\%$

5. $\frac{4}{16} = \frac{1}{4} = \frac{25}{100} = 25\%$

6. $\frac{8}{40} = \frac{1}{5} = \frac{20}{100} = 20\%$

7. $\frac{6}{20} = \frac{30}{100} = 30\%$

Page 59: Fractions Compatible with Common Percents

		Compatible Fraction	Estimate
1.	$\frac{21}{30}$	$\frac{2}{3}$	$66\frac{2}{3}\%$
2.	$\frac{5}{19}$	$\frac{1}{4}$	25%
3.	$\frac{16}{30}$	$\frac{1}{2}$	50%
4.	$\frac{7}{27}$	$\frac{1}{4}$	25%
5.	$\frac{4}{13}$	$\frac{1}{3}$	$33\frac{1}{3}\%$
6.	$\frac{3}{29}$	$\frac{1}{10}$	10%
7.	$\frac{8}{25}$	$\frac{1}{3}$	$33\frac{1}{3}\%$
8.	$\frac{5}{11}$	$\frac{1}{2}$	50%
9.	$\frac{10}{16}$	$\frac{2}{3}$	$66\frac{2}{3}\%$
10.	$\frac{4}{42}$	$\frac{1}{10}$	10%

Page 60: Estimate the Percent

		Compatible Fraction	Estimate
1.	$\frac{5}{49}$	$\frac{1}{10}$	10%
2.	$\frac{20}{31}$	$\frac{2}{3}$	$66\frac{2}{3}\%$
3.	$\frac{12}{25}$	$\frac{1}{2}$	50%
4.	$\frac{4}{15}$	$\frac{1}{4}$	25%
5.	$\frac{41}{62}$	$\frac{2}{3}$	$66\frac{2}{3}\%$
6.	$\frac{31}{42}$	$\frac{3}{4}$	75%

Page 61: Use Compatible Fractions

		Compatible Fraction	Estimate
1.	$\frac{4}{17}$	$\frac{1}{4}$	25%
2.	$\frac{3}{29}$	$\frac{1}{10}$	10%
3.	$\frac{11}{23}$	$\frac{1}{2}$	50%
4.	$\frac{7}{18}$	$\frac{1}{3}$	$33\frac{1}{3}\%$
5.	$\frac{9}{13}$	$\frac{3}{4}$	75%
6.	$\frac{71}{140}$	$\frac{1}{2}$	50%
7.	$\frac{9}{35}$	$\frac{1}{4}$	25%
8.	$\frac{1}{11}$	$\frac{1}{10} = 10\%$ lost	90% won

Page 62: Find the Total, or 100%

1. 10% of <u>40</u> is 4.
10% = 4
100% = 40

2. 20% of <u>35</u> is 7.
20% = 7
100% = 35

3. 25% of <u>160</u> is 40.
25% = 40
100% = 160

4. 50% of <u>180</u> is 90.
50% = 90
100% = 180

5. 5% of <u>160</u> is 8.
5% = 8
100% = 160

6. $33\frac{1}{3}\%$ of <u>30</u> is 10.
$33\frac{1}{3}\%$ = 10
100% = 30

7. 10% of <u>200</u> is 20.
10% = 20
100% = 200

8. 25% of <u>2,000</u> is 500.
25% = 500
100% = 2,000

Page 63: Mentally Find the Total

1. 60% of <u>25</u> is 15.
20% = 5
100% = 25

2. 40% of <u>200</u> is 80.
10% = 20
100% = 200

3. 70% of <u>70</u> is 49.
10% = 7
100% = 70

4. 90% of <u>30</u> is 27.
10% = 3
100% = 30

Page 63: Mentally Find the Total (continued)

5. $66\frac{2}{3}$% of <u>15</u> is 10.
$33\frac{1}{3}$% = 5
100% = 15

7. 4% of <u>500</u> is 20.
1% = 5
100% = 500

6. 30% of <u>90</u> is 27.
10% = 9
100% = 90

8. 75% of <u>32</u> is 24.
25% = 8
100% = 32

Page 64: Estimate to Find the Total

| $30 |

1. 50% of _?_ is $31.05
100% = $60

| $10 |

6. 20% of _?_ is $9.98
100% = $50

| $5 |

2. 10% of _?_ is $5.10
100% = $50

| $3 |

7. 5% of _?_ is $2.95
100% = $60

| $8 |

3. 40% of _?_ is $7.89
10% = $2
100% = $20

| $12 |

8. 3% of _?_ is $12.35
1% = $4
100% = $400

| $100 |

4. 25% of _?_ is $98.79
100% = $400

| $50 |

9. 10% of _?_ is $49.50
100% = $500

| $24 |

5. 75% of _?_ is $23.95
25% = $8
100% = $32

| 5% | | $20 |

10. 4.8% of _?_ is $19.75
5% = $20
100% = $400

Page 65: Estimate the Discounts

Estimates will vary. Exact amounts are given in parentheses.

1. About $200 ($197.19)
a) overestimate

4. About $8.00 ($8.25)
b) underestimate

2. About $6.00 ($5.78)
a) overestimate

5. About $300 ($324.75)
b) underestimate

3. About $3.00 ($2.92)
a) overestimate

6. About $40 ($37.60)
a) overestimate

Page 66: Estimate the Best Buy

1. Food Value has the better buy.

Food Value	Apple Foods
Save about: $.20	Save about: $.45
Estimated price: $3.80	Estimated price: $4.05

2. Wolverine Appliance has the better buy.

Wolverine Appliance	Sherman Appliance
Save about: $30	Save about: $100
Estimated price: $270	Estimated price: $300

Page 66: Estimate the Best Buy (continued)

3. Nelson Hardware has the better buy.

Nelson Hardware	Fred's Hardware
Save about: $15	Save about: $9
Estimated price: $15	Estimated price: $18

Page 67: Do You Have Enough Money?

1. no   **3.** yes   **5.** no
2. no   **4.** no   **6.** yes

Page 68: Review of Percent Estimation

Estimates will vary. Exact amounts are given in parentheses.

1. $6.00 ($5.79)
2. $6.00 ($6.44)
3. $3.00 ($2.83)
4. $500 ($493.25)
5. $70.00 ($69.50)
6. $20.00 ($20.55)
7. 25% (25.65%)
8. $30.00 ($29.45)